THE FAMILY LIFE OF THE MOSQUITO

A PRIMER FOR BIOLOGY STUDENTS, NATURALISTS, AND THE INQUISITIVE

GARY J. TORRISI PH.D.

The Family Life of the Mosquito by Gary J. Torrisi – First Edition

Copyright 2021 by Gary J. Torrisi

Printed in the United States of America

ISBN: 978-0-578-87910-9

Table of Contents

Preface

Why do I find mosquitoes so interesting and so important? Because mosquitoes are the most dangerous animals in the world. Mosquitoes also have the distinction of being the deadliest animal causing more deaths than the total loss from all the wars in the collective history of human existence.

It has been my intent from the onset of writing this book to bring to emerging biology students, knowledge accrued from the past to the present and eventually beyond, regarding the family of *Culicidae*. We all had to start somewhere, and this primer offers the reader the information needed to begin what has all the earmarks of a long journey.

It would be an overwhelming task to attempt to cover all of the information associated with the world of mosquitoes. Therefore I have targeted the undergraduate science majors, as well as interested naturalists, in offering the opportunity to delve into a portion of mosquito knowledge. To gain an appreciation of the volumes of information on mosquitoes, all you need to do is type on the subject line of your search engine the word 'mosquito, and, as in my case, 179,000,000 sites become available for perusal. The continual increase of manuscripts where mosquitoes are the most reported upon family of the insect world, will undoubtedly lend itself to revisions of our current understanding of the family of mosquitoes. As the readers maneuver through this book it is my hope that they will gain the same passion for the mosquito as I have.

The information in this primer may be used as a jumping off point by instructors to aid in course development and lay the groundwork for independent study for those students seeking to expand their knowledge of science and begin a career in scientific research. As a former teacher, I think about the possibilities where students may acquire interests into a study of the *Culicidae*, if, as the author, I can provide tantalizing facts that pique their interests.

A variety of subjects within the biological sciences lend themselves to a greater undestanding and appreciation of the role mosquitoes play in our everyday lives. It is not just for the entomologists, inclusive are the roles played by botonists, parasitologists, immunologists, ecologists, virologists, and by the medical and veterinary entomologists as we continue to battle the mosquito as transmitters of disease.

I have defined some of the technical terms used in this book. On the other hand, I have also, by intent, chosen not to explain nor define other terms. I believe that any student who has taken the time to read this book will already possess the inquisitive nature for self-discovery. The readers efforts to determine an understanding of terminology through searching the annuls of biology will promote a greater appreciation, a long-lasting understanding, and develop a thirst for knowledge stimulated throughout the chapters of this book.

It has been my experience that students need to grasp the importance of vast expanses of time, sequential data, and to recognize the value of merging evolutionary biology with the earth sciences. Therefore, chapter one delves into some of the history and progress related to changes in the earths landscape, climatic change, and the diversification of life that have defined the earth as we know it today. The geologic timetable, located at the end of chapter one provides the reader with a means by which they can begin to associate the chronological life changes to the earth's continually modifing landscape.

The reader will quickly realize that the chapters of this book are short by design. This allows you to go back and forth to re-read portions that seem pertinent to your needs.

Enjoy this work and may you benefit from its contents.

Acknowledgments

I would like to thank the following colleagues for the time they have set aside to review this manuscript. Their advice and productive criticisms have helped mold this book into a teaching tool for students and nascent naturalists.

My gratitude goes to Professor E. A. Heinrichs, Emeritus Professor, University of Nebraska-Lincoln. Dr. Heinrichs served as the Secretary-General of the International Association for the Plant Protection Scientists.

My appreciation also extends to Professor Wyatt W. Hoback, Oklahoma State University. Dr. Hoback has the distinction of being recognized in 2020 by the National Excellance in Teaching and Student Engagement. Excellance in College and University Teaching in Food and Agricultural Sciences. In 2017 he received the National Entomological Society of America Distinguished Achievment Award in Teaching.

Chapter 1: Comprehending Vast Time Periods

TIME and distance have always challenged our ability to percieve and con-
ceptulize. Take for example the speed of light and the vast distances covered
by that traveling wave. I can estimate a distance with some accuracy on a
golf course without the aid of a range finder but I am in awe when I try
to comprehend the vastness of distance measured in light years or parsecs,
and the speed of heavenly bodies. Consider our probes into space traveling
at 17,000 miles per second and the time it takes our probes to reach those
planets so relatively close to earth. Perception of these measurable forces
play upon our mental capacities. I cannot comprehend the distance or
speed of light from the sun to earth in eight minutes at 186,000 miles per
second covering 90 million miles. These factors just don't compute easily
like the calculated yardage I can quickly assess to hit a golf ball with a selected golf club. A million of anything is a challenge to our conceptualizing
based upon our compilation of information. I'll leave any other explanation of the cognizant mind to the psychologists to explain.

A working knowledge of the geologic time table will require repeated
referencing as you read through the evolutionary history of the mosquito.
As you work through this time-line of events the difficulty lies in our ability
to place into context a relationship of 'millions of years ago'. Our thinking
related to time is comprehended in a much shorter span of hours, days, or
even calender years. The geologic time-scale is measured in units referred to
as MYA (millions of years ago). It is often helpful to associate land and life
form changes over the 4.6 billion years of earth's existence. This coupling
places the concepts of time and life changes into an approachable understanding of the relationship between life forms and their changes over vast
expanses of time. Included in the figure are the moments in history where
massive extinctions occurred halting the change in the evolutionary adaptations of life. Dating techniques, fossil discoveries, tectonic studies, geologic formations, and characteristics of organisms help set the boundaries
presented in the table. The accumulation of these findings has contributed
to the developing evolution of the mosquito

An analogy that ties together the value of largeness, millions or billions of something, over lengthy time intervals can be depicted in this popular analogy. If you calculate the number of seconds that must pass to get
to one million accumulated seconds you will find that it is approximately
11 ½ days. Now make that super leap to one billion seconds of passing
time. What answer did you get? I offer the reader this mental gymnastics
because it is important to realize that evolution is rarely an observable event
when measured in human time scales.

The reader should find the timetable helpful in locating the placement of evolving early arthropods, the insects in general, and the mosquitoes place in the earth's changing history.

Fossils have a value of successive deposition chronologically where organisms succeed one another in a definite order. These fossils provide for an associated time scale in which the organisms lived in relation to the strata. There exists a compilation of Index Fossils that have proven to be a reliable source in the relative age of the earths rock formations and life histories. Index fossils are primarily aquatic remains of life during the Paleozoic Era. Extinction of those organisms is evidenced by their absence of fossil records during the Permian period around 250 MYA. The most notable of these fossils are the Trilobites, one of the earliest arthropoda, that lived during the Cambrian to the Permian periods.

By the way, one billion consecutive seconds requires approximately 31.5 years.

Eon	Era	Period	Epoch	MYA	Life Forms	North American Events
Phanerozoic	Cenozoic (CZ) / Tertiary (T)	Quaternary (Q)	Holocene (H)	0.01	Extinction of large mammals and birds / Modern humans (Age of Mammals)	Ice age glaciations; glacial outburst floods
			Pleistocene (PE)	2.6		Cascade volcanoes (W) / Linking of North and South America (Isthmus of Panama)
		Neogene (N)	Pliocene (PL)	5.3	Spread of grassy ecosystems	Columbia River Basalt eruptions (NW) / Basin and Range extension (W)
			Miocene (MI)	23.0		
		Paleogene (PG)	Oligocene (OL)	33.9		
			Eocene (E)	56.0		Laramide Orogeny ends (W)
			Paleocene (EP)	66.0	Early primates / —Mass extinction—	
	Mesozoic (MZ)	Cretaceous (K)		145.0	Placental mammals / Early flowering plants (Age of Reptiles)	Laramide Orogeny (W) / Western Interior Seaway (W) / Sevier Orogeny (W)
		Jurassic (J)		201.3	Dinosaurs diverse and abundant / Mass extinction	Nevadan Orogeny (W) / Elko Orogeny (W)
		Triassic (TR)		251.9	First dinosaurs; first mammals / Flying reptiles / —Mass extinction—	Breakup of Pangaea begins / Sonoma Orogeny (W)
	Paleozoic (PZ)	Permian (P)		298.9		Supercontinent Pangaea intact / Ouachita Orogeny (S)
		Pennsylvanian (PN)		323.2	Coal-forming swamps / Sharks abundant / First reptiles (Age of Amphibians)	Alleghany (Appalachian) Orogeny (E) / Ancestral Rocky Mountains (W)
		Mississippian (M)		358.9	Mass extinction	Antler Orogeny (W)
		Devonian (D)		419.2	First amphibians / First forests (evergreens) (Fishes)	Acadian Orogeny (E-NE)
		Silurian (S)		443.8	First land plants / Mass extinction	
		Ordovician (O)		485.4	Primitive fish / Trilobite maximum / Rise of corals (Marine Invertebrates)	Taconic Orogeny (E-NE) / Extensive oceans cover most of proto-North America (Laurentia)
		Cambrian (C)		541.0	Early shelled organisms	
Proterozoic		Precambrian (PC, W, X, Y, Z)			Complex multicelled organisms	Supercontinent rifted apart / Formation of early supercontinent / Grenville Orogeny (E)
				2500	Simple multicelled organisms	First iron deposits / Abundant carbonate rocks
Archean				4000	Early bacteria and algae (stromatolites)	Oldest known Earth rocks
Hadean				4600	Origin of life / Formation of the Earth	Formation of Earth's crust

Geological timescale with events. http://hyperlinks.phy/geological timetable
Precambrian (W,X,Y, and Z) represent organic sedimentary deposits.

CHAPTER 2: EVOLUTION AND DIVERSITY

Ape with Skull or Philosophizing Monkey designed by
Wolfgang Hugo Rheinhold in 1892.

"Nothing in Biology Makes Sense Except in the Light of Evolution."

THIS famous quote is not from some archaic philosopher centuries ago. This popularized statement was from an essay written by Theodosius Dobzhansky (1900 – 1975) and published in <u>The American Biology Teacher</u> in March of 1973. The message delineated in this quote shares the same lofty pedestal associated with Darwin's theory of evolution by natural selection and of the scientific method. The intent of this essay was a treatise critical of anti-evolutionist creationism and intelligent design that promoted theistic evolutionary ideas.

Ernst Mayr in his book <u>What Evolution Is</u> (2001) organized the subject into: evidence in support of evolution, what is meant by selection and adaptation, and why it leads to changes in the biodiversity of organisms. This chapter discusses the evolution of the mosquito (Diptera: *Culicidae*), provides examples of diversity through adaptation and selection, and offers examples of those adaptations displayed over millions of years.

Insect diversity is extremely valuable to humans by their actions of pollination and ecosystem services (dung removal, natural control of pest

5

species, and as members of all terrestrial and freshwater food webs). In fact, insect diversity has a great deal to do with promoting diversity of life. Surprisingly to many, the mosquito serves as a pollinator. However, most important is that the mosquito's ability to transmit disease is the leading cause of death and suffering for humans.

It is the mechanism of natural selection that promotes diversity of organisms through evolution of populations as they adapt to changing conditions. When a species attempts to colonize a habitat, it is the adaptations, or the fitness of the mosquito to adapt and reproduce in that habitat. These are the determinants that promotes the success or the death of the colonizing mosquito. Control of the process is reliant upon abiotic and biotic conditions selected for the establishment of a viable population able to develop, reproduce, and eventually succumb to their programed death.

The insect is the most diverse of invertebrate organisms and therefore provides science with valuable insight into evolution. Insects arose during the Silurian period about 420 MYA. However, the earliest of discoveries are not of the insects but rather the tracks left behind as an indicator of pioneer insect behavior during the early to mid-Ordovician several millions of years ahead of the Silurian period. Over time, several mosquito orders have gone extinct while many of our modern mosquito orders appeared by 250 MYA. The same can be said about modern families of mosquitoes that came into being during the Cretaceous around 120 MYA. What we know today related to the mosquito has its roots imbedded in fossil discoveries. Herein lies the problem of an incomplete record of existence. The lack of a consistent record applies to the same argument regarding the lack of evidence of the genus *Homo* and earlier vestiges that provide little clues. As time goes by, we are tantalized with new findings of human remains as science continues to build upon these discoveries. The same holds true for any organism, especially those extinct, including several mosquito genera. Since the early 1980's, hundreds of families and genera have become a welcome addition to the fossil record as more fossils of mosquitoes are unearthed from preserved strata and amber.

Insects have existed for about 350 to 450 million years and has taken their toll on the human population, birds, and other vertebrates. In comparison, human existence has occupied the same habitats for a little more than 2 million years. During this expanse of time, insects have adapted to life in most every type of habitat with the sole exception of the oceans and the continent of modern-day Antarctica. In addition, they have survived five great extinction periods. Over that expanse of time insects have developed amazing morphological characteristics, fascinating behaviors, and complex social interactions including the division of labor notable in bees, ants, and termites.

Fossil records of mosquitoes, although somewhat rare, have provided insight into the morphology of the mosquito and their biogeological lineage. Fossils of ancestral mosquitoes date back to the late Mesozoic era

(Jurassic and Cretaceous periods) and the Tertiary period (Eocene and Oligocene epochs). Edwards (1923) has gone so far as to suggest that earlier ancestral mosquitoes will be discovered pushing the lineage back to the early Mesozoic. There are no known mosquito fossils found from the Paleozoic era.

Difficulty lies in the understanding of taphonomy, the process by which fossils are formed. Limits to unearthing whole adult mosquito fossils and the noted absence of larvae or pupae in amber or sedimentary deposits complicate the deciphering process as researchers strive to apply a taxonomic hierarchy to an extinct genus of mosquito. In consideration of the small size and delicate body of the mosquito, conditions need to be near perfect for a mosquito to be preserved. That requires the specimen to have survived being devoured by a prey, thus, allowing for entrapment in the sap of a tree or covered in a fine sediment. Once captured in amber or imprinted in sedimentary deposits, it then becomes available to have been unearthed thousands or millions of years later. The exoskeleton, composed primarily of a flexible chitin and a more rigid sclerotin, are the ingredients that allow for mosquito fossil preservation in sedimentary deposits.

Recent publications have reported on an extinct species of mosquito found in Canadian Cretaceous amber, a Tertiary deposit dating from about 79 MYA and possibly an ancestral member of modern *Culicidae*. Two additional fossil species of *Culiseta* were unearthed from the Kishenehn Formation in Montana dating back to the Eocene epoch. Perhaps the earliest complete fossil mosquito was discovered in Burmese amber placing the age to be about 90 – 100 MYA. Older fossil records are expected to be discovered possibly dating back prior to the breakup of the supercontinent Pangea at the Mesozoic-Paleozoic boundary.

Careful examination of the evolutionary fossil record along with contemporary studies has provided valuable clues to emerging mosquito lineage, their adapted changes in structure, and behavior.

One of the most significant structures of all invertebrates is the development of wings. The pterygota or winged insects have been shown to coincide with the diversification of the insect during the Paleozoic. Flight has allowed the winged insect to move to more favorable habitats, escape predation, seek out partners, locate food sources, and establish breeding sites where other animals may perish due to unfavorable conditions or potential isolation from a mate because of their inability to move. During the Permian, giant dragonflies with wing spans of more than two feet were one of the early flying predators. Today, mosquito size and speed has not adapted to their physical world. While flying mammals and birds continued to develop, the mosquito relied heavily upon agility and efficiency of flight.

In addition to the development of wings, those that enter diapause because of unfavorable conditions during times of draught or harsh winter conditions produce an antifreeze that provides an extended protection until the weather becomes conducive for their survival. The northern Pitcher

Plant Mosquito (*Wyeomyia smithii*) freezes within the fluid contained in the carnivorous Purple Pitcher Plant (*Sarrecenia purpurea*) later to emerge in the third larval stage in late Spring once weather conditions improve.

West Nile virus is the most notable disease transmitted by the mosquito in the United States and North America in general. Countries in the tropics and subtropics remain inundated by the transmission of several serious illnesses by blood-feeding mosquitoes. Topping the list is the *Anopheles* mosquito, a carrier of malaria (*Plasmodium* spp.) The death toll throughout history due to malaria surpasses any other diseases known. The number of deaths has been expressed by demographers who have estimated the total number of humans born to be about 108 billion. Their projections suggest that nearly 50% have died from some form of disease transmitted by the mosquito. I share this information from an evolutionary standpoint. It has been suggested that mosquito success has relied upon the more stationary lifestyle of even the earliest of human gatherings where disease transmission from mosquito to human has adaptively increased and continues to do so even today.

Finally, the reproductive capability of insects allows for a large abundance of progeny. One of Darwin's notable observations includes the abundance of invertebrate progeny required to maintain a healthy population by overproduction of offspring. I find that, typical of biology, the number of fertile eggs produced by insects, varies from as little as one, in opposition to Darwin's postulate, to hundreds or even thousands. The number of eggs deposited by the mosquito may range from 200 to 300 eggs during oviposition. Several never survive. Perhaps the reasoning concluded by Darwin led to the explanation of the need to overproduce eggs as viewed during his acute moments of observation. The length of time needed to transform from egg to adult is often essential to maintain the life cycle of that insect. The avoidance of predators is increased by adapting shorter intervals of their development stages at a time when they are most vulnerable. Short developmental time spans are a critical process for many insects to ensure survival time prior to ovipositing. This adaptation occurs in the mosquito *Aedes triseriatus*, the Eastern Tree-Hole mosquito, whose development is often measured in a few days from egg to adulthood. It is also notable to point out those species of mosquitoes who get their start in vernal pools and have adapted to a shortened developmental time requirement. Consequently, as that water source begins to dry up, the mosquito has modified its development by accelerating their maturation time from egg to adulthood.

Climate change, especially in the Northern Hemisphere shows signs of disruption. The arctic biome has relatively few species of mosquitoes, the most important may well be those living in the shallow summer pools of water settled upon the permafrost. The tundra ecology and the relationship between pollinators and the emergence of flowers that rely on insects to transfer their pollen is showing signs of a disruption in emergence timing where the insect and the flower are no longer synchronous. Consequently, climatic warming trends have altered the relationship between pollinators emerging earlier and the developing flowers that provide life sustaining nectar for survival. The arctic tundra is not the only habitat in the world where phenology is being altered often resulting in devastating effects upon plant and animal life. The short summer emergence of the 'snow mosquito' (*Aedes communis*) is showing signs of a longer life, emerging earlier, developing faster, and surviving longer. If emergence occurs too early, then those plants that rely on these insects either exhibit diminishing returns due to non-pollinating adult mosquitoes or outright endangerment as to be extirpated or driven extinct over time. Typically, this mosquito can bury itself under the snowpack with the added survival ability to supercool, a process where excessive water and body fluids are lost prior to deep freezing to protect cells from bursting. This is not a unique process reserved just for the snow mosquito but one that has been selected in other insects as an adaptation to the harsh, long, winter season. Typically, when the snow mosquito emerges, there are literally billions that survive a short period raising havoc and death to caribou, birds, and other mammals. Data indicates a longer survival life span for this species because of the warming trend has exacerbated life in the tundra.

To re-construct the past, to understand mosquito history, to classify or re-classify, science has built an evolving understanding of the insect world. Coordinated efforts between fossils and the geologic time scale sets the phylogeny of the insect world. The mosquito has adapted and plays a major role in their affect upon humans.

CHAPTER 3: A BRIEF LOOK INTO THE NUMBERS

Mosquito Taxonomy:
    Kingdom: Animalia
        Phylum: Arthropoda
            Class: Insecta
                Order: Diptera
                    Family: *Culicidae*

SPECIES diversity, richness, and abundance rely upon natural selection to garner a stable and reproducing population. Selection is made for those individuals that are 'most fit', able to reproduce, and repeat their life history. In the case of immigration and colonization of species, the greater the ratio between species and area yield a richer, more abundant diversity. This assumes little competition for food, mating contact, and niche allocation. Environmental variables like latitude, and altitude influence richness of a species. It is quite clear that as species move to northern latitudes or higher elevations their numbers decrease or in some cases remain stable at a lower abundance than what would be found in the tropics. Loss of richness of a species or the decrease in abundance has been noted through studies associated with human activity such as logging (Brazil and Indonesia), urban sprawl, and man-made fractured habitats that have occurred over relatively short periods of time. Natural events that occur on geologic time are a much slower process where habitat loss has the chance to recover or create new habitat based upon land and sea changes. Climatic change involving atmospheric gas ratios also have had a great effect. These events have taken place over several hundreds, thousands, and millions of years involving landscape change through coastal erosion, riverine flow, volcanic activity, and archipelago formations that have provided for new habitats inviting species immigration and colonization.

Harbach has listed the mosquito inventory of the world to be 3583 species as of March 2021. The numbers may change, up or down, rarely by more than one or two species. Additions to this list are related to the discovery of new species while deletions are often found to have been synonyms of an existing species.

The number of genera within the *Culicidae* family appears to be 113 worldwide. Changes in genera may occur where taxonomists/systematists offer supportive information to elevate a subgenus to the level of genus and vice versa. I am unaware of any new genera being recognized at this time. One of the most notable change occurred in 2000 when the subgenus *Ochlerotatus* was promoted to genus. This altered dramatically the number of recognized species in the genus *Aedes* and may have contributed to some of the confusion associated with the new ranking amongst researchers. A few years later *Ochlerotatus* was returned to a subgenus level re-establishing species back to the *Aedes* genus.

According to the CDC, within the continental United States and its territories the best estimate of the number of mosquito species is anywhere from 185 to 200 or more species. Of these, three genera hold the distinction of harboring the greatest number of pathogens and parasites while serving as carriers and transmitters of mosquito borne diseases. The three genera are represented by the *Aedes, Anopheles*, and *Culex* mosquitoes. West Nile virus is the most important of the diseases transmitted in the United States. In the first half of the 20th century, malaria was the most important of illnesses primarily introduced by travelers re-entering the states. Several of our more southern states are on the receiving end of invasive mosquitoes. The latest report of a recent invasive mosquito occurred in Florida and identified as *Aedes scapularis*. The more notable would be *Aedes albopictus* and *Aedes japonicus* both capable of transmitting illness. The very first *Ae. japonicus* specimens I captured were collected in a pine forest at the Albany Pine Bush, Albany, New York using an EVS trap with gas released CO_2 and Octenol as attractants. It should be noted that the type of trap used, the life stage desired, habitat selection, temperature, and time of year, to name a few variables, often dictates the species of mosquito collected. Therefore, a limit on the extent of the collection can be a limitation of the richness collected during that surveillance period.

Habitat certainly plays a major role in the diversity of mosquitoes in different states. As an example, West Virginia has approximately 35 species of mosquitoes. This may well be the lowest number of mosquito species of all the states except for Alaska. Of course, West Virginia, is small in area, and has limited habitat choices when compared to larger states that contain a greater variety of habitats. West Virginia is primarily a mountainous state where colder altitudes place a limitation on the establishment of breeding populations of mosquitoes. Mosquitoes that display a cold hardiness adapt to the West Virginia landscape while so many others probably die out during winter months and unable to establish a viable population the following year. On the other hand, if you were to select the state with the largest number of mosquito species Florida would stand out. A large state in area, warm climate year- round, subtropical in much of the state, and an abundance of shallow, warm, slow moving water offer advantageous conditions for successful breeding. Well, you would be wrong. It appears that Texas, by its size, variation of habitat's, and warm southern regions supports about 85 species of mosquitoes slightly more than Florida with about 80 species. I know, I am probably splitting hairs right now.

The number of species associated within a specific region, a country or state, will vary based upon reports from multiple sources. Trying to determine the exact number of species is more difficult than it appears. There are times where the number of species occupying a habitat escapes verification. It seems, far too often, that there is a lack of reporting or sharing of known richness of mosquitoes obtained during surveillance studies. I spent two years trying to determine the number of species in New York

State. With the understanding that wildlife does not adhere to manmade borders viewed on a map complicates matters as to what is a reproducing population or what species is a visitor from a similar habitat from across a state border. Due to the uncertainty, I have suggested to colleague's that at least two consecutive years of capture from the same location is necessary to verify that a reproducing population be counted as an occupant of the state. At best, I have only been able to establish the number of mosquito species in New York State to be about 63. I determined the number of species using the collection housed at the state museum in Albany, and from information provided through recent published studies including my capture of *Ae. japonicus*. However, according to the New York State Department of Health webpage, about 70 species are reported. I am probably missing one or more of the species as there have been no recent collections with new species offered and therefore absent from the museum's archives. Perhaps, some species have not been captured for a long period of time.

As of this writing there are 13 recognized genera in the United States. New York had 9 of these 13 identified as reproducing populations. Recently I came across a surveillance study from 2018 centered around New York City. Listed in their work were captures of *Toxorhynchitis*. The Albany museum is an old and diversified collection of mosquito specimens dating back to 1879. However, not one *Toxorhynchitis* specimen existed in the vast collection coming from New York until this report verified a substantial collection of *Toxorhynchitis* over a period of two or more years. That brings the total of genera in New York State to 10. They include, *Aedes, Anopheles, Psorophora, Culex, Culiseta, Coquillettidia, Orthopodomyia, Uranotania* and *Wyeomyia*, along with *Toxorhynchitis*. Three other genera are found throughout the United States but not in New York State. They include *Deinocerites, Haemagogus,* and *Mansonia*. Located at the end of this chapter I have provided a listing of mosquito species archived at the museum or identified and reported through publications or personal capture that exist in New York State.

Heterogeneity, habitats with multiple availability of niche occupancy, has a greater richness of species than those habitats with fewer occupiable niches. I refer the reader to recall the differences in mosquito richness across several states based upon the number of habitat choices. Often, co-existence of species occurs within a habitat by partitioning of resources. An example of partitioning occurs in the carnivorous northern Purple Pitcher Plant, *Sarracenia purpurea*, which is often occupied by three obligate insect species. Inclusive are the mosquito, *Wyeomyia smithii*, a midge, *Metriocnemus knabi*, and a flesh fly, *Fletcherimyia fletcheri*. In each case, they separate themselves distally and temporally based upon respiratory requirements, foraging habits, and the time of year they occupy and exit the leaf. All three insects are obligate dipterans and must live part or all their early life stages in the fluid of the plant. Co-existence occurs by mutualistic partitioning within the plant. The mosquito larvae swim to the air-water interface to fulfill its oxygen requirement. Free swimming larvae feeds throughout the column on the inquiline organisms that flourish in the plant fluid and the larvae remain within the boundaries of the plant fluid until it reaches adulthood. The midge wormlike larvae forages for nutrients at the bottom of the detritus filled fluid column, obtains oxygen through an abdominal bladder organ, and is the first to arrive in the spring. The flesh fly larvae, the last to arrive, often only one larva is deposited, rests on top of the water due to its respiratory restrictions, feeds on captures at the surface of the plant fluid and crawls out of the plant leaf within 10 to14 days.

So, what are the numbers that reflect diversity and abundance of a species. This is a very dynamic process. It takes seasons and even years to establish relevant numbers of a species within a designated area, compare their consistency of populations and if not repeatable, why? What habitat changes might be the reason for lack of a population and determine any change in phylogeny of the species. The big picture is bleak. Loss of habitat through anthropomorphic practices and the resultant climatic changes have disrupted the ecology of the world. Dailey, we read about the loss of diversity in some of the most disrupted habitats. It is this simple. If there is a continual destruction of habitat and human behavior continues to increase the levels of carbon dioxide and methane into the atmosphere, at a rapid pace that exceeds any known historical times, then loss of species and the loss of interaction between species is the ultimate endpoint. Yes, I realize that there are some species that can increase their abundance and extend their range through change. I live in a suburban area surrounded by a fractured mixed temperate forest. I have observed an increase in ants, termites, aphids, grey squirrels, and chipmunks. The deer no longer cross my property. I have not seen a bluebird in summer or a titmouse for several winters. I was pleased to observe a small flock has returned to my yard during the 2020-2021 winter season along with a flock of the black-capped chickadee and the slate-colored junco. We can all recognize that change is inevitable, but, not at the fast pace we are observing.

Mosquitoe of New York State

Ae. absesratus	Ae. trivitatus	Tx r. septentrionalis*
Ae. albopictus*	Ae. vexans	
Ae. atropalpus		Ur. lowii
Ae. aurifer	An. barberi	Ur. sapphirina
Ae. canadensis	An. crucians	
Ae. cantator	An. earlei	Wy mitchelli (?)
Ae. cinereus	An. punctipennis	Wy. smithii
Ae. communis	An. quadrimaculatus	
Ae. decticus	An. walkeri	**These archived specimens**
Ae. diantaeus		**were not captured in NYS.**
Ae. durifer	Cq. perturbans	Ae. aegypti (?)
Ae. dorsalis		Ae. atlanticus
Ae. excrusians	Cx. erraticus	Ae. flavescens
Ae. fitchii	Cx. pipiens	Ae. mitchellae
Ae. fulvus pallens (?)	Cx. restuans	Ae. spenceri
Ae. grossbecki	Cx. salinarius	
Ae. hendersoni	Cx. territans	An. bradleyi
Ae. implicatus		
Ae. intrudens	Cs. impatiens	Cs. annulata
Ae. japonicus	Cs. incidens	
Ae. melanura	Cs. inornata	Ps. mathesoni
Ae. moristans	Cs. melanua	
Ae. provocans	Cs. moristans	
Ae. punctor	Cs. particeps	
Ae. restuans	Cs. silvestris (minnesotae)	
Ae. riparius		
Ae. sollicitans	Or. alba	
Ae. sticticus	Or. signifera	
Ae. stimulans		
Ae. taeniorhynchus	Ps. ciliata	
Ae. thibaulti	Ps. columbiae	
Ae. triseriatus	Ps. ferox	

(?) unlikely breeding in New York

CHAPTER 4: HOLOMETABOLIC LIFE CYCLE

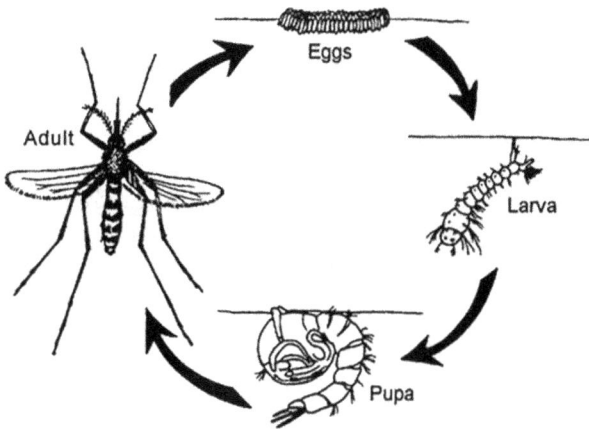

Mosquito Life Cycle

THE ability of insects to succeed as a species in the environment is determined by the adaptive forms and functions of the species driven by the process of natural selection.

Metamorphosis, or the changes that occur during developmental stages of an insect are not all the same. In general, there are three forms of development. Ametabolic, where eggs develop into tiny adults and continue to grow until they become sexually mature. This form of metamorphosis includes some of the more ancient forms of insects that include springtails (*Collembola*) and silvertails (*Thysanura*). Incomplete (heterometabolic) metamorphosis involves development of egg to nymph to adult. The nymph is the growing stage where wings are externally developed. There is generally no quiescent stage, a non-feeding stage, referred to as the pupal stage. Examples of heterometabolic insects are grasshoppers (*Orthoptera*), mantids (*Mantodea*), cockroaches (*Blattodea*) and termites (*Isoptera*). Complete (holometabolic) metamorphosis has four life stages of development. Egg, larvae, pupa, and adult. The larvae have no resemblance to the adult. Larvae continue to feed and grow while it may also reside in unrelated habitats and require different foods. Most insects are holometabolic including the mosquito (Diptera: *Culicidae*). Several holometabolic insect larvae have unique names. Examples would be for the butterflies and moths referenced as caterpillars, flies as maggots, beetles as grubs, and mosquitoes as wrigglers. Of course, there are exceptions to the general descriptions presented here of holometabolic stages of some insects.

Climatic conditions, especially temperature swings, draught, and winter conditions have a direct effect upon the development period of insects. Harsh conditions force insects into a life pausing dormancy referred to as diapause. Diapause may occur during any of the stages of complete metamorphic development. Some insects may require more than one season to complete their development, while others, develop in a single one- year cycle. Univoltine is the term used for a one-year cycle that provides a single generation. There are bivoltine insects that produce two batches of progeny in a single year as well as multivoltine insects that usually do not require a period of diapause unless severe climate conditions delay (draught or unusually extreme temperature swings) development of the species. My favorite example is the Pitcher Plant Mosquito (*Wyeomyia smithii*) located in the northern latitudes that enter diapause in the third larval stage in late summer and remains within the frozen water of the plant over the winter months only to emerge in late spring and advance their development to adulthood. This univoltine species requires overwintering to satisfy the required adaptation to protect and guarantee the propagation of the species.

The mosquito life cycle is holometabolic or complete metamorphosis and displays four unique life stages. These stages include the egg, four larval instars, a non-feeding pupal stage, and the adult. All stages are aquatic except for the adult. Furthermore, there are no similarities between the larvae and the adult, they live in different habitats, forage on a different food source, and at the end of a non-foraging stage, the pupa, the adult then emerges.

THE EGG:

EGGS (IN EGG RAFT)

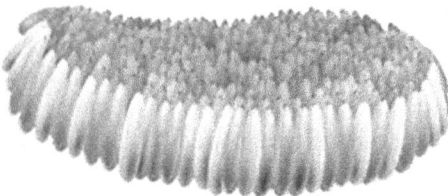

Typically, mosquito eggs are soft, pliable, and white. Within a short period of time, depending upon the species, the egg will harden and darken to a brown or black coloration. At the time of oviposition, eggs are permeable and may experience water loss. Hatching of the egg occurs once the embryo is fully developed. Hatching may be triggered by low concentrations of oxygen, temperature sensitivity, circadian rhythm, and possibly the introduction of bacteria.

The time of year, a preferred ambient temperature range, and a myriad of other cues leads the gravid female to either accept or reject the conditions that controls the selection of a satisfactory oviposition site. The female, depending upon the species, may oviposit anywhere from a few dozen to hundreds of eggs often not all in the same container or pool of water. The color of the container, the size, height above the ground, existence of conspecifics, density, and dozens of other cues direct the female as to whether eggs will be deposited. This process displays a complexity of heterogenous factors.

Eggs are generally coated in a protected shell originally soft and flexible but later becomes hardened and waterproof. The shell allows for the transfer of atmospheric gases while conserving water loss. The embryo continues to form within the egg, while some species, those that are drought resistant, can delay the hatch until conditions become favorable (*Aedini*). Some are capable of survival for months in these harsh conditions. *Aedini* species lay eggs where rain or high tide is necessary to stimulate egg hatch that is delayed or may not occur for some time. When it does, an apparent population explosion of mosquitoes is experienced.

Eggs are deposited in clusters known as egg rafts onto the surface of shallow, slow moving, or quiescent water pools and float until the egg hatches. This is typical of the genus *Culex*. Other genera known to lay eggs in rafts include *Culiseta*, *Coquillettidia*, and *Uranotaenia*. The *Anopheles* female deposits her eggs, one at a time, directly upon the surface of the water. The *Aedes* female will generally oviposit her eggs, one at a time, on the side of a natural or artificial container and hatch once the water level rises and triggers the hatch releasing the first larval stage. *Mansonia* deposit their eggs in a rosette cluster upon aquatic vegetation under the surface of the water.

Embryonic development begins almost immediately after the eggs have been deposited. This development may take only a couple of days to a week or more. Once the first instar larvae are fully developed it hatches from the egg and requires a watery environment to survive, feed, grow and molt. The eggs of *Culex*, *Culiseta*, *Coquillettidia* and *Uranotania* are arranged vertically within the egg raft, the larvae are developed with their anterior end pointing downward toward the water and hatch directly into it.

THE LARVAE

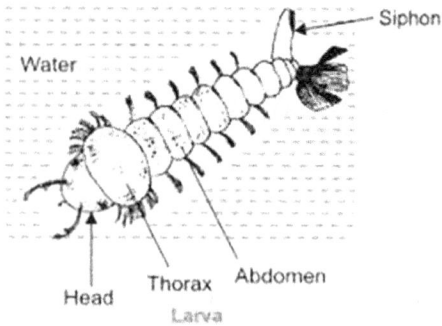

Mosquito larvae are often found living within the structures of water filled cavities of plant parts. Phytotelmata is the term associated with these habitats. Living quarters for mosquito larvae are often temporary or permanent shallow pools of water that have little to no movement or confined to various sized artificial or natural containers. There have been anecdotal reports of mosquito larva observed in a bottle cap as well as the more extreme size where larvae occupy several hectares of marshlands and swamps. Most larvae live in fresh water with few exceptions that are found within brackish water. *Aedes sollicitans* and *Aedes taeniorhynchis*, the Salt Marsh mosquito and the Black Salt Marsh mosquito are two examples of mosquitoes that breed in brackish water. Obligates like *Wyeomyia smithii* live within the pitchers of the pitcher plant where they remain from egg to pupa.

When the egg hatches, the resulting first instar larvae is already prepared for the watery life that is maintained through four instars, known as 'wrigglers', and the pupal stage, referred to as 'tumblers', prior to the emergence of the terrestrial living adult. Survival and development rely upon a watery environment, their ability to rise to the air-water interface to use the atmospheric oxygen that enters through open spiracles located at the end of the respiratory siphon.

In addition, an inquiline complement of microscopic food particles, algae, bacteria, and other single celled microorganisms are present to forage upon.

The lack of water movement is unnecessary to bring food particulates to the feeding larvae. The head is equipped with moving brushes that sweep particles into the awaiting mouth. Additional feeding for some species is the added ability to scrape the sides and bottom of the container they live within. *Toxorhynchites*, however, is a predator of invertebrates, often other mosquito larvae. This species is quite large when compared to other larvae

18

species and equally the largest species as an adult. Adults of the species feed on nectar and are not blood-feeders or predatory foragers.

The end of the larval abdomen contains siphons that break the air-water barrier and take in the needed oxygen from the air. The air travels through a network of tubes called trachea that run systemically throughout the larval body. Although differences occur between the subfamilies of *Culicinae*, *Toxorhynchitinae* and *Anophelinae* species, two genera, the *Mansonia* and *Coquillettidia* are quite unique. These mosquito larvae remain submerged and receive their oxygen by cutting their way through the underwater stem of plants. This behavior provides access to the hollow tubing of the plant stem allowing the larvae to remain submerged.

Toward the final stage of development undifferentiated cells eventually develop into adult structures. It is in this stage that 'imaginal disks' develop quickly and are responsible for the formation of the adult appendages prior to the emergence of the pupa.

THE PUPA:

The pupa remains as an aquatic, non-feeding stage of mosquito development. The pupa consists of a fused head and thorax with an abdomen supporting two appendages called paddles. The paddles are used as a locomotive structure and when disturbed it tumbles down from the surface, an escape mechanism. Hence the term 'tumbler' describes the avoidance reaction of the pupa.

Respiration occurs through thoracic spiracles which are surrounded by two 're-spiratory trumpets' that protrude from the dorsal mesothorax out into the surface air. The pupa generally remains motionless at the surface where their trumpets take in oxygen laden air. Much of the adult organs are formed from undifferentiated embryonic cells. The heart and fat body carry over into the forming adult. Once the adult is formed within the pupa, the pupa lies at the air-water interface where it begins to gulp air.

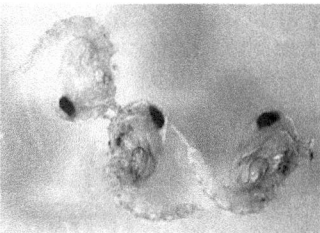

The increased pressure from the air causes the cuticle to rupture and the adult releases from the pupa casing and rests upon the water surface or a nearby landform. This process, if temperatures are high enough, only takes a day or two to complete this final stage of development.

THE ADULT:

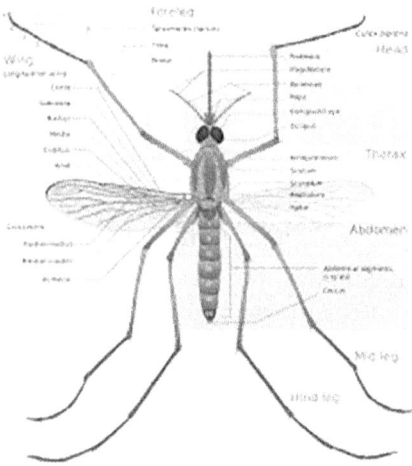

Once the adult emerges from the pupal stage, the adults generally seek shelter within vegetation, and small hollow shaped caverns, their resting habitat, only to leave the protective surroundings when they become active. Typically mosquitoes will take flight once or twice a day based upon an internal circadian rhythm that is triggered by the light-dark cycle. Mosquito flight activity may be described as diurnal, nocturnal, or crepuscular (dawn and dusk cycles). These flight times are often associated with foraging patterns, mate seeking, or locating oviposition sites. Habitats vary depending upon their needs. Some are forest dwellers while others prefer open areas. Mosquitoes will generally travel less than 2 km in their lifetime. Flight speed is difficult to measure but it is known that as headwinds inhibit flight speed, the mosquitoes will tend to avoid flight unless it is synchronized with a strong tailwind. Mating usually occurs within a day or two of the hatches. The males form swarms over swarm markers, objects that contrast with the environment, where each male performs a looping flight pattern over the marker in hopes of attracting a mate. When a female enters the swarm, the males can distinguish the female by her species characteristic wing beat sensed through the male's antennae and Johnston's Organ.

The adult mosquito is comprised of three body parts or tagmata. The head is equipped with five appendages. These appendages include two antennae, the primary chemical sensing organs, which especially sense body odors such as lactic acid and the main attractant to female mosquitoes, carbon dioxide. The pedicel located at the base of the antennae contains the Johnston's Organ. These organs serve as mechanical receptors that sense vibrations of sound. The sounds may serve as a defensive warning signal, or it may pick up the wing beats of a mate. Two palpi that also serve as sensory organs are located next to but inside the antennae. The proboscis, the feeding apparatus is located near the middle of the other two pairs of appendages. The proboscis serves as a nectar feeding organ and as a skin piercing blood-feeding organ. It comes equipped with two tubes within the sheath of the proboscis. One tube injects saliva containing an anti-coagulant and a pain killer. The anti-coagulant keeps the blood from coagulating long enough to draw its meal while the pain suppressant does exactly that.

That is why you may experience the after effect of the 'bite' (although mosquitoes do not bite). The proboscis repetitively probes for a suitably sized vessel then begins to draw blood. The second tube draws up the blood along with an excess of water and salts. The mosquito can take up to 3 to 4 times its own body weight in blood and must discharge the excess fluid and salts rapidly. The weight of the fluid hinders flight and the discharge of these fluids and salts for some species occurs while they are still feeding. The dorsal side of the head has two eyes capable of detecting the slightest movement while two smaller eyes (ocelli) detect changes in light intensity. However, once feeding begins, the female usually does not fly away even after sensing a shadow approaching from overhead.

The thorax contains several long appendages. These appendages are comprised of three sets of long paired legs and a pair of elongated wings. In some species, the banding and location of those bands on the legs are important identifying characteristics. In addition, the second pair of wings have been modified into a small club-like structure called halteres. These paired halteres aid the mosquito in flight orientation. Therefore, the wings lend translation of the order Diptera described as the two-wing order. There are two sets of spiracles on the lateral sides of the thorax, which are sometimes associated with or without setae used in identification. The spiracles are part of the trachea system that delivers oxygenated air to the body cells.

The elongated abdomen is comprised of ten segments where usually only 8 are visible. They are often banded and are characteristics that lend themselves to identification features. The dorsal side are termed the terga, and the ventral side of the abdomen, the sterna. All terga except 1,9, and 10 house a pair of spiracles tied directly to the tracheal system. The digestive, excretory and reproductive systems are found in the abdomen.

Many of the features discussed in this chapter can be viewed with the aid of a stereoscope. External morphological structures are of value to anyone wishing to identify the mosquito species. Banding or the lack there of on the proboscis, thorax, or the abdomen are characteristics used for identification along with wing patterns. The *Anophelines* provide a clear example of the location and number of wing spots used to identify the species within the genus. In addition, the location of banding, basally or apically, serve a defining characteristic in identification. Setal location as pre-spiracular or post-spiracular on the lateral thoracic provide valuable identifying characters. Finally, sclerotized plates on the head and thorax also serve as identifiable external morphological structures.

It takes patience and practice to master the art and science of mosquito identification. A good identification book along with knowledge of the region and time of year the mosquito was collected become important aspects of the identification of mosquito species. There is more to identifying insects than what I have shared. If you put in the time required for collecting, pointing, and maintain sound records, over time you will have broadened your expertise in mosquito identification.

CHAPTER 5: NECTAR AND BLOOD

Nectar feeding *Toxorhynchitis*; note the curved proboscis that eliminates the ability to blood feed.

WE have all experienced being probed by a mosquito in its need to obtain a blood meal. However, the only blood-feeders are the female mosquitoes. In consideration of the 3500 plus mosquito species known, only a few hundred blood-feed. I am not aware that a specific number of mosquito species that blood-feed has ever been definitively determined. All mosquitoes, male and female alike, feed on nectar as their necessary energy source. Upon emergence mosquitoes will nectar feed. In some species, the female, driven ultimately by the adapted genetic control, will seek out a vertebrate host to blood-feed. This may occur as early as 2 to 3 days while other species may seek a blood meal immediately after eclosure. Few female mosquitoes will skip the initial nectar feeding prior to seeking their preferred vertebrate host. The female uses the nutrient rich blood for the maturation of her eggs while somatic cell repair and replacement, as well as energy needs, are obtained primarily through foraging for sugar rich plant nectar. The blood meal does not seem to play a large part in the energy procuring process primarily due to the low availability of amino acids in nectar required for ovary maturation. However, there is some evidence that the blood may serve as a low energy source for the female in addition to the energy supplied through nectar foraging of both genders. If it was a necessity to blood-feed for maturation of gametes, cellular repair, or replacement, then all mosquitoes, non-blood-feeding females as well as male mosquitoes, would have evolved into blood-feeding organisms. Over time, the female anatomy and physiology has developed to accommodate the blood-feeding females. It is of interest to note that in some females, just prior to diapause, blood- feeding behavior ceases, and storage of nutrients gained from nectar is deposited and stored in the fat body becoming available after emergence from diapause. Blood provides the necessary nutritional based components (amino acids) for the female mosquito while the nectar provides the needed energy for sustainable living.

If you were asked what the most important requirement was for mosquitoes that provides for their survival and reproductive success, it would be food. Longevity and reproductive success are related to their food source or lack thereof. Some mosquitoes are opportunistic foragers often termed generalists. Others are more explicit in their selection process and are called specialists. This is not a hard and fast rule. If plant or vertebrate hosts are not available, the obligate mosquito will be forced to take its meals from a non-specific host.

Nectar feeding by the mosquito relies on their olfactory, visionary, and taste senses. Sources of sucrose rich nectar are found primarily in the plant's flowers. Other sources of energy rich food include extra-floral nectar, honeydew, and rotting fruit. Since males do not blood-feed, going without a source of nectar leads to their death. Climate change, specifically increase in temperatures, causes some insects to emerge several days ahead of their preferred plant host that has yet to flower and is likely to cause a disruption in their phenology. Early emergence that takes place ahead of flowering of their plant preference interferes with their survival and interferes with pollination by the insects. Nectar feeding is a function of species differences, habitats, seasonality, and volatile organic compounds.

Volatile organic compounds are produced by plants in a gaseous form. These carbon-based compounds serve plants as pollinator attractants as well as a defensive volatile chemical designed to ward off herbivores and parasites without chasing away nectar feeders. This is often referenced as an 'evolutionary arms race' that continually develops between plants, and, as in this case, insects. The plant produces a new volatile chemical weapon, and the herbivore adapts to the change and counters the novel chemical. Mosquitoes rely upon the odor produced by the flower, a group of chemicals called terpenes. Detection of terpenes occurs through the sensilla of the mosquito's antennae.

Blood foraging upon vertebrates has evolved over millions of generations and has been adapted by several female mosquito species. This adaptation has lent itself toward maximizing its reproductive fitness. Host selection responds to a myriad of cues that may be sensed as an attractant, or perhaps adversely, and a resultant rejection of that host by the mosquito. Several cues have also been identified as attractants or rejection cues during female oviposition site selection. Odorants, primarily carbon dioxide, contributes to mosquito selection of a vertebrate host. However, a lack of host availability, perhaps due to adverse weather conditions, seasonality, anthropomorphic regulation, temperature, and humidity all influence the flight range of the mosquito. Energy reserves are taxed, and a secondary source of blood may be selected if cues are determined to be detrimental to their reproductive success. Changes in seasons is known to cause certain species to switch hosts. For example, *Culex nigripalpus* also switches from feeding on the blood of deer in the summer to birds in the winter. The significance of mosquito blood-feeding is directly tied to the transmitting of disease in humans and domestic animals. Did you ever think about the increase in the human population over thousands of years as compared to mosquito existence dating back millions of years? Has the population boom of human life provided select mosquito species an abundance of accessible and specific hosts leading to the emergence of disease in humans?

The adapted feeding structure of mosquitoes, the proboscis, is the sole organ- system responsible for accessing nectar and blood. The proboscis is one of the five appendages attached to the head of the mosquito and is modified to pierce and draw blood or nectar. It is this organ that provides for the act of insertion into the skin, pump saliva into the puncture and draws the needed blood from its vertebrate. This process must occur quickly, within minutes, as it serves as a protective behavior for the survival interest of the female. The probing that takes place involves quick and repetitive movement to seek out a vessel larger than capillaries. Once the vessel, a venule or arteriole is located, saliva is released delivering an anti-coagulant and an anesthetic. The anti-coagulant prevents clotting of the blood in support of the feeding behavior. The anesthetic often keeps the host unaware of the invasive piercing by numbing the area causing recognition of a recent bite after the mosquito completes its blood taking and has departed. The saliva delivers antigens causing a reaction or localized bump at the site of the injection. If the mosquito saliva harbors a virus or other pathogen, it is injected along with the saliva early in the process. Soon after piercing of the proboscis, a second pumping system draws in the blood at a volume of 2 to 6 ml depending upon the mosquito species. In some species, diuresis allows for the discharge of excess fluid to reduce the weight of the mosquito so as not to interfere with a burdensome flight. In some species it will release a clear fluid while other species may have a reddish or deeper red tinge appearing fluid. The release of fluids can take place while the female is feeding, so, once again, the mosquito is defecating on your skin.

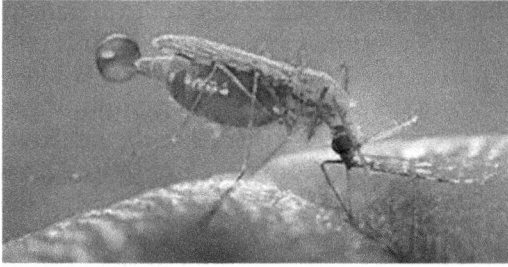

The proboscis has an enveloping sheath known as the labium that folds back exposing the two pumps and the piercing stylets or cutting tools. In addition, the proboscis is equipped with neuromuscular tissues and chemical sensing setae called sensilla. Some parts are sclerotized while other structures remain softer and pliable. The structures enveloped within the labium are the piercing, saliva injecting, blood drawing apparatus collectively referred to as the fascicle. The female uses the same mouthparts for both blood-feeding and nectar feeding. Food sources can be located several meters away through chemically sensitive organs located on the tarsi and palpi. Most notable is the recognition of carbon dioxide released by vertebrates during respiratory activity and has been measured to be sensed by female mosquitoes several meters from the vertebrate host.

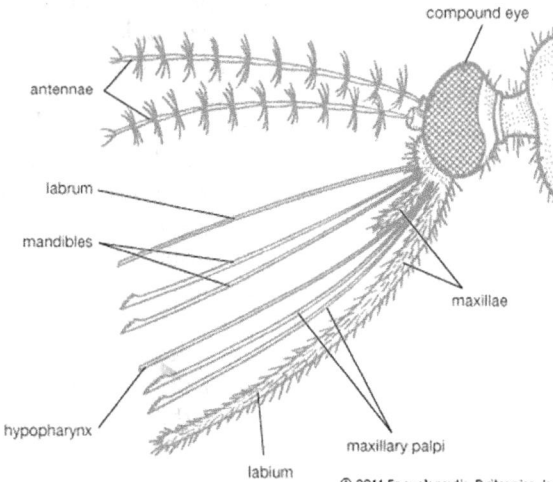

Mouthparts of a mosquito

compound eye

antennae

labrum

mandibles

maxillae

hypopharynx

maxillary palpi

labium

© 2011 Encyclopædia Britannica, Inc.

The fascicle may repeatedly probe the skin for the larger blood vessels and may fly away early in the process when sensing host movement. This often occurs during the initial phase of insect exploratory behavior where the female may remain motionless for a few seconds prior to initializing probing and penetration of the skin. Probing for a satisfactory vessel seems to rely upon chance and may be the reason for multiple probing within the epidermis. Once attained, saliva is released at the penetration point. Once blood drawing occurs, the mosquito does not seem to react to movement of the host. I am sure you have swatted an engorging female leaving behind a blood smear upon your skin. If the female is left alone the process is then terminated in response to the swelling of the abdomen of the female mosquito.

The host, humans, receive one or more exposures to the proteins in the saliva of the feeding mosquito. Immune response by the host is common and subsequent bites usually reveal one of two allergic reactions to those bites. The first immediate reaction is termed Type-I Hypersensitivity, an inflammation of the skin centered around the fascicle probing area. This is also referred to as wheal-and-flare where the allergic response starts within minutes and may last a few hours. The second delayed allergic reaction, Type-IV Hypersensitivity involves a cellular immunity reaction by lymphokines that released by the antigen triggered T cells. In both cases, the result is itching, redness, and slight swelling around the site.

The mosquito's ability to locate and ingest plant nectar or vertebrate blood begins with the uniquely adapted mouth structure, the proboscis. The need to convert foods into usable chemical energy and to provide the necessary amino acids for reproductive success requires digestion of the foods. Nutritional requirements may be processed and used immediately or delayed for later needs. The digestive process begins with the salivary juices and continues through the digestive track where the bulk of the digestion occurs in the midgut. Following the breakdown of nutrients in the midgut, the end products are then absorbed and delivered to the hemolymph for life sustaining needs and reproductive success. Meals of nectar taken by the female are stored in the crop and is absorbed in the midgut at a relatively slow pace leaving the stomach empty and ready to take on a blood meal when needed. Digestion of blood continues to be reduced into simpler chemical components throughout the digestive system where waste, excess heat, and fluids are expelled through the anus. This final process occurs in the hindgut portion of the digestive track along with the osmotic regulation of the Malpighian tubules. During the processing of blood, the protein compounds are broken down and make available important amino acids to be reconstituted for their needs that are utilized by the female in the maturation of her eggs.

A final note about feeding reveals that insect feeding strategies are often either plant or animal sources or both. Food may be either in the liquid (adults) or solid form (larvae). The resultant digestive systems in insects have evolved over the countless of generations through some unique and effective processing of foods. The mosquito, especially the female, forages on both plant material and vertebrate blood resulting in its own structural adaptation within this family of insects.

Completion of a blood meal requires the involvement of the endocrine regulatory system. Maturation is a result of regulation, as outlined by Briegel (2003) who stated that following eclosion, anatomical differences, behavior maturity, hormonal changes, the developing digestive system, vitellogenesis and even flight all must be developed prior to blood feeding.

Have you ever considered the regulatory mechanisms that control your ingestion of a meal and the regulatory mechanisms required in digestion and excretion to obtain healthful nutritional value and a continuous energy supply? Food for thought!

CHAPTER 6: OVIPOSITION SITE SELECTION

Natural Tree-hole

THE previous chapter discussed the mechanics of acquiring food by mosquitoes. I referenced the energy requirements of flight muscles obtained from nectar foraging and the blood-feeding by females required for egg maturation. Once the nutrient requirements are satisfied, egg deposition can take place. The process of selecting a suitable and safe site follows a series of steps prior to depositing her eggs. These steps include courting a mate, copulation, and the storage of sperm in the female organ called the spermatheca. The final step in the process is fertilization that occurs in the oviduct of the female and those fertilized eggs are stored until the decision is made to oviposit. Interestingly, males emerge and swarm a day or two ahead of the eclosion of the adult females. This is referred to as protandry. It has been suggested that early emergence of males is a 'double-edge sword' for male mosquitoes. On the one hand, the earliest of male emergence gives that male the upper hand in female courting and copulation and may have the superior genetic make up to pass along to offspring. On the other hand, males become targets for predators if they fail in their hunt for a suitable mate.

I became interested in egg deposition by pure accident. I was aware of the repetition in literature that suggested females select dark tree hole containers, often black or brown, whether they be natural or artificial and the selection of plant structures such as bracts and pitcher plants with lighter color attractants for select mosquito species. Some species will deposit eggs in shallow, quiescent waters. Vernal pools come to mind. *Culex* mosquitoes will utilize both quiescent pools and tree-hole containers suggesting that site selection is not limited to either microhabitat. My accidental discovery took place on a bright, sun filled day in the back of my warehouse. A white, four by eight- foot set of artificial stairs were leaning back against a fence. The risers and treads formed a V- shape providing a holding area for water to collect. In the shallow water, no more than 6 inches deep, I observed each V section to be teeming with wrigglers. A few days later I observed a similar pattern in a white flowerpot half filled with white sand holding several inches of water containing several mosquito larvae. These observations seemed contrary to what I had been reading. Intrigued by these observations I was determined to research this further. I formed my hypothesis, designed, and conducted my research, and later published my results. During this time, an exhaustive literature search of dozens of published manuscripts revealed a myriad of cues that mosquitoes use as either attractants or deterrents in selecting a particular deposition site.

Oviposition site selection, in addition to the quiet waters of vernal pools previously mentioned also include natural or artificial tree-holes. Water-bearers or phytotelma, bromeliads, bracts, and my favorite, the carnivorous pitcher plant all serve as a microhabitat selected by the gravid female to deposit her batch of eggs. Even damp, leaf covered, forest floors serve as a niche for the deposition of some *Aedes* species. As a reminder, all life stages, excluding the adult, require water to continue their development. Tires are the most referenced artificial container. Importation of tires from Asia was the vehicle that brought us the invasive Asian Tiger mosquito, *Aedes albopictus*. In addition, common water holding structures around the house include bird baths, house gutters, flowerpots, and untreated swimming pools. To control mosquito outbreaks in and around your home, the simplest treatment is to drain all unnecessary artificial water holding structures to reduce breeding of mosquitoes.

In selection of a site for deposition of eggs by a gravid female, she will decide from available cues which is deemed an attractive niche or is a deterrent to the selection process. The range of travel is limited for most mosquitoes and therefore flight outside their range of activity is pre-determined and therefore the species remains within the confines of their more recent ancestral home. Tropical biomes display the largest diversity of mosquito species and provides for the greatest selection of habitat availability. As you move toward the poles, biodiversity drops off while site selection is reduced in availability. Temperate regions, followed by the Taiga or boreal forests, and finally the furthest northern zone, the Tundra biome, display a continual reduction in mosquito diversity and the reduction of site selection coupled with a drop in mean temperature due to seasonality. To be able to compensate, some mosquitoes enter diapause to preserve the existence of a population and is an adaptation that has evolved with the changing climates over millions of years.

In addition to color, several other cues are used to select a site for egg deposition. For each site selection made, the prospect of a safe niche must be made for the attractive value to insure survival and the fitness of their progeny. A choice made that is detrimental to the survival of offspring is avoided, while the correct assortment of cues provides for success of those offspring and is incorporated into the selection process. Of course, we do not know what that assortment is. Research often compares one cue against another. However, there often are conflicting results when compilation of numerous summations is compared as a metacommunity behavior. There is a continual competition for limited resources, both intraspecific as well as interspecific. Final selection influences the fitness of her progeny and the promotion of population stability.

Research takes place in laboratories, artificial settings like containers or cages, wild types located in natural settings, or by comparative settings. Mosquito research targets specific global habitats, selected mosquito species either singularly or through comparative application of two or more species. Life stage of the mosquito is the heart of research to conduct meaningful summaries of the researcher's findings. Application of these results may provide clues to mosquito abatement especially those known to transmit disease. Examples of site selection cues abound in the literature. Here are some examples.

In Panama artificial tree-holes of four different colors were set up in unfiltered light of open gaps on the forest floor and under-story light filtered canopies that reduced or blocked sunlight. Gravid females overwhelmingly selected the under-cover canopy over the open, sunlit locations. The author (Yanoviak, 2001) suggested that the sunlight may have an influence on the perception of color in their selection process. Yanoviak (1999) also investigated detritus quality and quantity and the affect it might have on the productivity and distribution of the mosquitoes by comparing four litter types. The effects of leaf litter composition among those four plants suggested preferential selection measured by the comparative totals of richness and longevity of *Culex mullis*. Species preference for leaf litter and the effect on their productivity and distribution in natural tree-holes species has merit.

Competitive displacement of one species over another was studied by Brakes (Brakes et al. 2004) in Brazil. Shared habitats and the resultant change in distribution of two species showed that an established population can outcompete the other species for the same habitat. In the United States, the invasive mosquito *Aedes albopictus* often displaced previously established populations of *Aedes aegypti*. In this case, *Ae. albopictus* was the superior competitor of the two species vying for the same habitat. It is important to note that this superior advantage is not uniform and *Ae. aegypti* often remained the dominant species when in direct competition with *Ae. albopictus*.

Frank (1986) tested plant form and color choice as an influence of site selection in bromeliads under a laboratory setting. Monitoring the choice of mosquitoes in the genus *Wyeomyia* involved shaded areas, distinction of leaf like form, and visual perception of color. Visual perception received consideration because the mosquitoes in this study are daytime flyers. Site selection involves visualizing color, form, and sunlit areas versus shaded areas. The plant revealed a flower spike that seemed to be an added attractant, however, as I have previously alluded to, earlier research suggested that it may serve as a distraction and reduce selection.

Toxorhynchitis is the largest of mosquitoes. The females are non-blood-feeders, and their larvae are known to be predators of other tree-hole mosquito species. Jones and Schrieber (1994), in a laboratory setting, selected eight colors that were placed at ground level and at one meter above the ground. Two *Toxorhynchitis* species, *Tx. splendens* and *Tx. rutilus*, were monitored as to color choice and ground level selection versus color selection and height preference. *Tx. splendens* primarily selected the black containers followed by dark red and brown. Over 90% of their eggs were deposited at ground level. *Tx. rutilus* also selected black containers as their preferential choice along with red, blue, and brown. Twenty-five percent of their eggs were deposited at the one-meter height. This study suggests that ground removal of discarded containers and artificial water filled containers serve to control mosquitoes.

Mosquito density as it relates to oviposition preferences and the abundance of predators was conducted in New Zealand (Lester & Pike 2003) by monitoring the behavior of *Culex pervigilans*. They found that an increase in density that occurred in the smaller surface area containers and depth of water was not significant from a statistical viewpoint. However, in larger surface area containers with a greater depth of water was preferential to predators.

The primary tropical African malaria mosquito, *Anopheles gambiae*, was the focal point of a study in Kenya (Minakawa et al. 2005). A relationship between habitat stability and the occurrence of *An. gambia* seeking indoor resting sites and the availability of resting areas was tested. There was a positive correlation between habitat size and a continuation of water availability over time. They found that habitat size is the determining factor for habitat stability and resulted in adult abundance when pupae were monitored.

Onyabe and Rottberg (1997) conducted a laboratory investigation involving the mosquito species *Aedes togoi*. The researchers looked at the presence of conspecifics as to whether they would serve as an attractive cue for the gravid female or would the presence of conspecific eggs lead the female to avoid that container. Avoidance of the presence of conspecific eggs by the female was viewed as a protection to avoid intraspecific competition that could limit the abundance and health of the larvae. When waste material, bacterial growth and the addition of food supplements was added, increased oviposition occurred when compared to rearing in water alone. In addition, *Ae. togoi*, like that of *Ae. aegypti*, select more than one site to deposit their eggs. This behavior may be interpreted to increasing the probability of success by not keeping all her eggs in one container (or basket).

There are several other investigations into the site selection process utilized by gravid mosquitoes. It is important to highlight that site selection influences the fitness of mosquito larvae and this corresponds to the selection of positive cues and the rejection of unfavorable ones. These determinations are a result of the females use of her olfactory and gustatory sensory organs. Are the plant infusions and micro community within the containers suitable for the larval development? Do conspecifics serve as attractants or are they to be avoided? What about predation occurring after the eggs have been deposited? There is the problem of infection, can the female detect lurking pathogens that may harm or eliminate her batch? There is the problem of strong competition for food, space availability and its effect upon density. What about drought, remembering that all stages of a mosquito's life, except for the adult, requires water. My research suggested that under an exceptional snow-less winter, a dry spring and sparse early summer rains, the conditions delayed ovipositing until ample rainfall filled the leaves of the plant. The number of first stage larvae, often found to be over two dozen in a leaf occurred in late July. These first stage larvae, usually observed a month earlier, are now required to forage and molt over a shortened period of the remaining summer to reach the diapausing third stage no later than September.

Life histories of mosquito species are not all the same. Generally, traits may be grouped as they relate to reproduction and development or provide escape from adversity through dormancy or even migration. A lingering question revolves around the phenology of the mosquito and its adaptations to climate change.

CHAPTER 7: A HISTORY OF INFLUENCE

If you think you are too small to make a difference,
try sleeping with a mosquito.

14_{th} *Dalai Lama*

SEVERAL years ago, I received a book authored by Speilman and D'Antonio entitled <u>Mosquito: The Story of Man's Deadliest Foe</u>. Part one describes the mosquito as a 'Magnificent Enemy'. What a great choice of words to kick off a web of entangled history, disease, and death. The writers offer countless stories that played a role in influencing the outcome of human history. The very first page of the preface sums up the power and importance of the mosquito.

> *"Mosquitoes have felled great leaders, decimated armies,*
> *and decided the fates of nations."*

What a better way to capture the intrigue proffered by the role played by the mosquito. I became interested in discovering who, when, and the end results that were influenced by diseases transmitted by the blood-feeding of the mosquito.

Poet and politician Lord Byron (1788-1824) described the effect of illness transmitted by mosquitoes as "early adult mortality". Lord Byron died at the age of 34 from malaria.

Going through a list of notable individuals I often am asked, how do we know, for example, that Alexander the Great perished from malaria in 323 B.C.? Great question considering the antiquity of medical treatments practiced prior to modern medicine. In some cases, co-morbidity was the deciding factor leading to death. If fortunate, physicians' records or at least some anecdotal writings recorded the conditions of the individual prior to their demise. In the case of Alexander the Great, historians have speculated that he may have been exposed to a combination of typhus, malaria, alcohol poisoning and perhaps even Guillain-Barre' Syndrome or GBS for short. GBS is the result of several forms of illness including infections from bacteria and encephalitic types of viruses like Zika. In some instances, knowing where and when notable persons were at the time lends clarity in determining their illness or death. Another case of co-morbidity was likely the cause of death of Osceola (1804-1838). Osceola was a leader of the Seminole's during the second Seminole Wars. Contrary to popular belief, he was not a Seminole chief and was not even a member of the Seminole tribe. He was born Billy Powell to a Creek mother with Scottish, English, and African mix on his paternal side. He was born in Alabama and was raised by his Creek mother in the culture of her native American society.

He suffered from chronic malaria and acute tonsilitis known as quinsy, an infection that spreads beyond the tonsils to surrounding tissue. A weakened immune system brought on by two illnesses in Osceola's case is what co-morbidity reflects on the ill-health or demise of an individual. Covid-19 reflects the very same nature of co-morbidity issues and often leads to mortality within a population during a pandemic.

Famous military leaders that either became chronically ill or died from malaria often included their armies and led to the defeat or retreat of invading forces. Attila the Hun's (452 AD) army was stopped in Rome due to malaria. Otto I attacked Rome in 964 AD and lost almost his entire forces not from battle but from malaria. On December 7, 983, his son Otto II died of malaria at the age of 28. Genghis Khan (1162-1227) chose not to invade Western Europe due to malaria outbreaks affecting his troops. Here is another case of uncertainty where historical accounts offer a mixed message suggesting blame be placed upon his falling from his horse, struck by an arrow in the knee, or from malaria.

Early notables infected prior to the use of quinine in the 17th century include Dante in 1321 and King Edward IV in 1483 who died from a co-morbidity which included malaria. Columbus, during his second voyage was infected, along with his entire crew from malaria on Hispaniola. After 1631, the use of quinine to treat malaria was introduced into Europe. The papacy had begun construction of a new home in 1574 because so many of the inhabitants of Rome including several Popes were sickened or died from malaria. Malaria in parts of Italy was so infectious that a summer retreat into the hills above the swamps became standard procedure. This practice took them away from disease causing mosquitoes to a higher, cooler home. At the time there was no correlation between malaria and mosquitoes but during WWII Hitler had these very same swamps that surrounded Rome, flooded. The thought was to induce increases in mosquito populations knowing that malaria would slow the advancement of the allied troops.

The Continental Congress, during the early stages of the American Revolution, appropriated $300.00 to purchase quinine for our troops. During the war, entire British garrisons were infected by malaria and it has been speculated by some historians that it was the toll taken by the severity of malarial infection that led, in part, to the British surrender at Yorktown.

Least of all, let us not forget the results of yellow fever infection or the combined effect of yellow fever and malaria. Napoleon's army in 1801 had to abandon their effort to quell uprisings in Haiti. In addition, the financial losses incurred from this campaign led to the Louisiana Purchase years later.

The delays due to illness and death along with the cost over-ride in building both the Suez Canal and the Panama Canal were a direct result of both yellow fever and malaria epidemics. These conditions forced the French to give up on the building of the Panama Canal. Soon after the project was undertaken by the United States. To quell the effects of yellow fever and to protect and reduce the effects of the illness upon workers, several doctors including Walter Reed were dispatched to Cuba. It was through a system of isolation from mosquitoes when compared to the unprotected, a correlation between mosquitoes and yellow fever transmission was established for the first time.

War has certainly taken its toll upon military personnel often changing the course of events and the outcome of battles. During our Civil War, 1,316,000 soldiers were infected with malaria and somewhere around 10,000 perished. Estimates of the time suggest 50% of white soldiers and 80% of black soldiers had experienced relapses each year.

In 1895, the French were involved in a war in Madagascar. Thirteen troops died in battle while over 4,000 died from malaria. In WWI, French, British, and German soldiers were unable to fight for a period of 3 years due to repetitive malaria outbreaks. General Douglas MacArthur, the commander of our forces in the south Pacific during WWII had available only one-third of his troops at any given time during the Pacific campaign. The other two-thirds were hospitalized from malaria or in recovery and deemed unable to perform their duties as soldiers. During the Korean War, 629 of our U.S. military contracted malaria each week. The war in Vietnam recorded 40,000 cases. Interestingly, we never knew the degree of illness that had played a part in limiting the enemy during the history of United States warfare.

Several of our presidents contracted malaria. Washington was 17 years old. Monroe had several reoccurrences having been infected in Mexico around 1785. Jackson, who spent time in Florida during the Seminole Wars was infected. Lincoln in his early years contracted malaria. Other presidents known to have been inflicted were Grant, Garfield at the age of 16, Teddy Roosevelt in 1914 during a trip to Brazil, and finally Kennedy, during his WWII tour of duty in the south Pacific.

A list of notable figures throughout history have contracted malaria and some perished from its effect. Authors, actors, athletes, researchers, political leaders, and even some of my childhood heroes are noted in the history books. I believe the reader can grasp the significance of vector-borne illness transmitted by the tiny insect known as a mosquito.

CHAPTER 8: ARTHROPODS AND DISEASE

THERE exist over 500 known viruses that are in association with arthropods. Of these, less than half are known to be carriers or even transmitted by mosquitoes, while about 100 are known to infect humans. Later chapters will discuss some of these diseases related to three families and four genera of viruses and are arboviruses. They are the *Togaviridae* (genus *Alphavirus*), *Flaviviridae* (genus *Flavivirus*), and the *Bunyaviridae* (genera *Orthobunyaviridae and Phlebovirus*).

The phylum Arthropoda is the largest group of invertebrates which includes the insects, the largest class within this phylum. The name refers to 'jointed -legs' and includes trilobites (all extinct), crustaceans, millipedes and centipedes, arachnids (ticks and mites), and the insects. Parasites, bacteria, and viruses are transmitted by several representatives of the arthropods. Several diseases may also be transmitted by other means. Ingestion of the pathogen by swallowing infected crustaceans, poor hygiene practice, contamination by frass deposits, crushing the insect on your skin, and spreading the infection all contribute to illnesses. Vector-borne diseases affect hundreds of millions of people worldwide where many of these diseases may be preventable. Vector-borne diseases effect 17% of all infectious diseases causing nearly one million deaths each year. Malaria transmitted by the *Anopheline* mosquito causes 400,000 to 500,000 deaths each year with a yearly estimate of over 200 million cases worldwide. Most deaths occur in children under the age of 5 years. Dengue is the most widespread viral infection transmitted by the *Aedes* mosquitoes. Estimates suggest 3.9 billion in 129 countries are at risk of infection with an estimate of 96 million symptomatic cases resulting in approximately 40,000 deaths per year. Asymptomatic cases become the hidden factor of contracted disease cases globally. However, mosquito-borne illnesses are not horizontally transmitted between humans.

The genera or species of a competent vector determines the pathogen passed on to humans, domestic animals, and wild animals. There is a complex demographic set of determinants including social and environmental factors, unplanned urbanization lacking long-term strategies, as well as intercontinental flight of travelers often involving trade relationships. Viruses, for example, may extend their range through the expansion of their host range or by movement of the inflicted through travel aided most often by intercontinental flight. To be infected with African sleeping sickness, one needs to be bitten by a *Glossina* species of fly, the tsetse fly, that carries the protozoan *Trypanosoma brucei*. The fly is limited in range to sub-Sahara Africa and has not been introduced into other continents. The illness known as leishmaniasis, another protozoan, is transmitted through the bite of a sand fly of the genus *Phlebotomus*. This is an Old-World Disease and has spread into the southern part of Europe. The black fly of the genus *Simulium* transmits a nematode throughout parts of tropical Africa causing River Blindness. The disease requires several bites over time and may not present itself symptomatically for one or more years. The result of vector-borne illnesses from blood-feeding mosquitoes occurs primarily but not in total in the tropical and sub-tropical habitats. Further, there is a disproportion affect upon the poor populations within these countries which in turn stresses the health facilities and adds economical losses of man-hours to the region or country.

To broaden our thinking let us consider the effects that often accompany neglected tropical diseases (NTD) in poor populations. The World Health Organization (WHO) has addressed the problems associated with NTDs and the self-imposed suppressed help sought by the afflicted people who display a reluctance to speak about their condition. Individuals are stigmatized and ostracized, bullied, discriminated against, and have mental health problems. WHO has taken the global leadership in addressing five NTDs including lymphatic filariasis transmitted by the bite from a mosquito. WHO has reported that for the fifth straight year, more than one billion people have been treated through programs that prevent NTDs. Control and elimination on a large-scale preventive treatment campaign using volunteers and health-care workers are committed to improving the health standards of lives. WHO currently is combating 20 NTDs, contribute to the suffering of people, especially the poor and those that go without medical aide. The pandemic (Covid-19) has disrupted the progress of these programs creating shortages of health-care workers and health supplies. Pharmaceutical companies have for decades donated medicines including covering significant associated costs. In 2019, 2.7 billion pills were delivered to needy people of which 2.1 billion pills were managed by WHO. Their programs and the involvement of eleven pharmaceutical companies has attracted several donors to aid in the support of WHO's endeavor. It is my intention here to have the reader grasp the significance of mosquito diseases and the difficulty in testing, treating, and paying for the cost of medicines, hospital care, and the stress upon doctors, nurses, medical institutions, and the individuals and families afflicted. When I wrote this last sentence, I reflected upon my thoughts and felt that it is not necessary to remind the reader that the pandemic we face is quite clear and unnecessary to reiterate the obvious. Apparently my third thoughts prevailed.

Illnesses due to bacteria are not associated with bacteria-mosquito transmission. However, several insects are responsible for diseases such as typhus and trench fever (lice), tularemia (Diptera: *Chrysops* spp.), and most notable throughout human history, the plague transmitted by fleas.

Here in the United States we have been battling the effects of diseases transmitted through the bite of ticks and mosquitoes. Although malaria and dengue are global problems, infections of Lyme and other tick related illnesses are currently our biggest problem. Lyme disease has surpassed the collective number of illnesses compared to West Nile virus where both diseases are known primarily to have emerged in the northeastern U.S. and has since spread across the continent.

Three genera of mosquitoes are the main contributors to disease in humans. They are *Culex*, *Aedes*, and the *Anopheline* mosquitoes. Some of the important global illnesses transmitted by mosquitoes are filariasis, malaria, and a host of viral infections responsible for dozens of diseases. Once infected, the individual may be asymptomatic and totally unaware of the infection as is often the case of Zika, some strains of dengue, and West Nile virus. Individuals may be mildly infected with the ability to 'tough it out' or seek and receive treatment for symptoms. Often rare but not to be dismissed severe reaction to the disease may lead to the death of an individual who might ignore the more severe symptoms, not seek medical attention, or may simply not have access to treatment. Nonetheless, mosquito transmitted disease infects millions of people each year and kills over a million people during that time span. Although malaria is responsible for the bulk of the deaths and illnesses reported it is dengue that is far more widespread where cases are increasing each year on a global basis. Several of the viral illnesses, along with malaria and lymphatic filariasis are discussed in the following chapters.

Vector-borne diseases are difficult to control. The only vaccine currently available is for yellow fever. There is a novel approach to a potential malaria vaccine that I will address later. The CDC reported that from 2004-2016 nine vector-borne diseases were identified for the first time in the United States and U. S. territories. Control of mosquitoes and the animal reservoirs and hosts associated with viral pathogens is a disturbing challenge. The pathogens and the increase in abundance of mosquitoes and human contact ratios increasing in some habitats is a result of climate and environmental change. In addition, insecticide and antibiotic resistance is on the increase coupled with the adaptation of the pathogens to human immune response. Collectively vector-borne illnesses are far from under control and have led to several local epidemics and wide-ranging outbreaks especially in poorer nations with little or no medical facilities that can reach the afflicted. Illness from mosquitoes and other arthropods is a large and growing health problem in the United States. The CDC summarizes the implications for public health and to effectively reduce transmissions requires improvements in our surveillance, diagnostics, reporting, tracing, development of vaccines, and therapeutic medicines to go along with mosquito control strategies.

CHAPTER 9: FILARIASIS

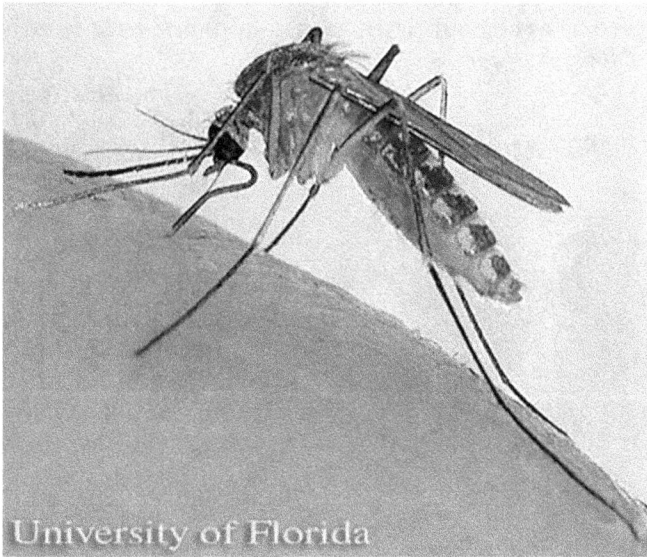

Photo: Female *Culex quinquefasciatus*, the Southern House mosquito, taken by James Newman at the University of Florida.
Vector of West Nile virus and filariasis.

PREVIOUS chapters have mentioned three important categorical diseases transmitted by mosquitoes that have a deleterious effect upon humans globally. They are the myriads of viruses, malaria, and the neglected disease of human filariases by parasitic nematodes.

Nematodes are the most abundant animal found throughout terrestrial and aquatic habitats. Experts note that they comprise 75 to 80 % of all animal life while some suggest this is a gross underestimate of the 15 to 20,000 known nematode species that exist. Soil nematodes are free-living, predominately terrestrial while parasitic nematodes are common in plants and animals. It is the parasitic roundworm in animals that will be discussed as they relate to two of the more common ailments in the United States and throughout tropical and subtropical countries. These worms are spread through the blood-feeding of mosquitoes and in the case of human infections, are the leading cause of permanent disability second only to leprosy. The human parasitic roundworms are threadlike and often microscopic although they may reach a length of about one foot. The head and tail are indistinguishable to the naked eye.

You may be familiar with hookworm, pinworm, trichinosis, and a research favorite of geneticists, *Caenorbaditis elegans*, referred to as

C. elegans a free swimming, transparent roundworm about 1 mm in length. This chapter will discuss two infectious nematodes, the first, *Dirofilaria immitus*, the dog heartworm and lymphatic filariasis, commonly referred to as elephantiasis, caused by *Wuchereria bancrofti*, *Brugia malayi*, and *Brugia timori*.

Heartworm Disease

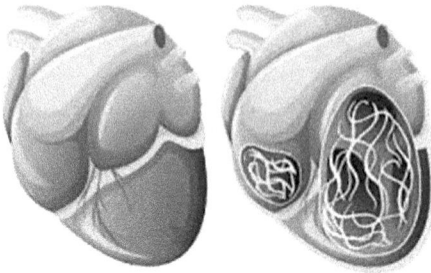

Dirofilaria immitus, the nematode associated with dog heartworm was first described in Italy in 1626 and later in the United States in 1847. Dogs are not the only mammals capable of acquiring this disease. Cats and ferrets along with fox and coyotes, both serve as reservoirs for the mosquito, and it has even been found in sea lions. There are approximately 25 species of mosquitoes capable of transmitting the parasite to mammals. The adults may live up to 5 to 7 years within the host. If untreated, it may be life threatening to your pet. Humans are rarely infected. The roundworm is not passed from dog to dog and requires a mosquito bite to introduce the larvae to the host mammal. Even though it is called "heartworm", the parasite invades the lung blood vessels and later migrates against the blood flow and lodges in the right side of the heart. The cycle continues as the adult releases its larvae into the blood stream of its host and is picked up once again through the proboscis of the mosquito apparatus upon blood-feeding. On a positive note, veterinary medicine has researched this disease and preventive measures are readily available including treatment of pets that show positive results when tested for the parasite. The drawback, as the adult parasites are killed through chemical treatment, those remains within the vessels of the lungs and heart may block those vessels and can be lethal.

Lymphatic filariasis is a neglected tropical disease where adult parasites reside in the human lymphatic vessels and nodes. Results of infection are impairment to the lymphatic system, abnormal enlargement of body parts, including the genitalia, resulting in severe disability and social stigmatization. Lymphatic filariasis is the second leading infection that results in severe disabilities after leprosy. The adult worms develop and can release as many as

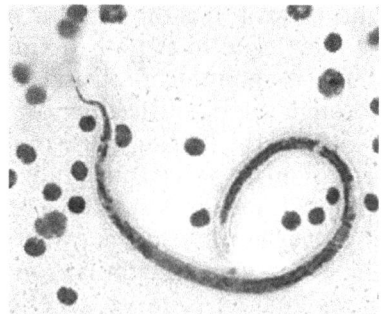

50,000 microfilaria each day and are introduced into the mosquito as third stage larvae microfilaria by blood-feeding on infected hosts. Adult parasitic nematodes may live up to 5 to 8 years and are capable of reproducing offspring. The infection in humans requires several bites over months to years before it manifests itself to a full-blown infection within the human host. The primary roundworm responsible for as much as 90% of all cases is *Wuchereria bancrofti* found in urban areas of the tropics, subtropics, and parts of Asia, Africa, Western Pacific Island nations, South America, and the Caribbean. It is no longer found in the United States where the disease disappeared in the 1930s. Rural areas of the south and southeastern Asia are targeted by *Brugia malayi*. *Brugia timori* is restricted to a few Indonesian Islands of the Pacific. Transmission of filaria is associated with several mosquito genera and over a hundred species of mosquitoes.

Interestingly, these parasites display two varieties of behavior known as nocturnal periodicity and nocturnal sub-periodicity. Nocturnal periodic types remain in the blood vessels of the lungs during the day and are not available to daytime feeding mosquitoes. In the evening, the microfilaria migrates to peripheral vessels and the lymphatic system. The night feeding mosquitoes pick up the microfilaria during that time. It is unknown if there are animal reservoirs that serve as intermediate hosts for the parasite, so it appears that mosquito to human is the mode of transmission. This behavior may be partly responsible for the longer life of the adult parasites. Nocturnal sub-periodic types do not follow the same regimented migration diel and may be ingested at most anytime of the day or night.

Tracing the evolution of roundworms is limited. There are no substantial written records known prior to the 16th century and historically speaking, cannot be confirmed. There are some ancient references to the presence of lymphatic filariasis dating back to 2000 BC and up to 500 AD. Symptoms associated with filariasis was first documented in 1588 from a reliable documentation of severe lymphatic filariasis symptoms denoting the swelling of the leg and foot. In 1863, and a few years later, the presence of microfilariae was noted from observations of fluid discharge in urine and blood extracted from the infected persons. It was around this time that an association was made linking microfilariae and elephantiasis. About a decade later, Joseph Bancroft observed and documented the first adult worms. The worm was later named after Bancroft as a new species known today as *Wuchereria bancrofti*. A year later, Patrick Mason discovered the microfilaria in mosquitoes. This significant relationship was later applied to other mosquitoes as transmitters of yellow fever and malaria.

Maturation of the nematodes and the modes of transmission to humans follow pretty much the same sequence of development. Once the mosquito ingests an infected blood meal, the infective third stage larvae travel toward the midgut of the mosquito where some loss occurs from inherent barriers of the mosquito digestive system. In a short period of some 30 minutes the microfilaria pass through the gut wall and enter the open haemocoel or body cavity containing the hemolymph. The parasitic larvae then travel to the thoracic muscles and then finds its way to the proboscis. This is now the infective stage that followed the incubation period of approximately 10 days from ingestion to infective mode. Note that there is no reproduction that occurs in the mosquito. When the mosquito takes its next blood meal, the larvae are released upon the skin of the human host. Most of the larvae die at this time, however, those that locate the puncture wound provided by the piercing proboscis then crawl through the opening. At this point in time, the nematodes molt twice, and then migrates to the lymphatic system usually in the lower extremities. Once within the vessels and nodes of the lymphatic system, the worms develop into adults. This takes somewhere between 6 to 9 months before microfilaria are detected in the blood. Note that the salivary glands are not involved in this process unlike viruses, bacteria, and protozoans that are pumped into the punctured skin. Furthermore, there is no reproductive cycle of the parasite within the mosquito. Human infections generally require contact from several repeated blood-feedings on the part of the infected mosquito.

To enact effective control measures, transmission of the infection requires knowledge of several key factors. These factors include, knowing the prevalence of the infected number of people, the density of microfilaria in the blood of the infected, the number and location of the transmitting mosquito, and the frequency of human mosquito contact in the infection zone. Since the adult worms are entangled in a mass within the lymph vessels and ducts, surgery to remove the mass is too intrusive and is an alternative to following mass drug administration (MDA) to eliminate the microfilaria but is limited in its effect upon the adult worms. MDA is a strategy, along with vector control to eliminate illness from communities around the world. In 2000, the WHO established the Global Program to Eliminate Lymphatic Filariasis (GPELF). Initially, 81 countries were targeted as in need for intervention. Recently, 16 of these countries have eliminated the parasite while 7 others are on the brink of doing so. It remains a problem even with the success already noted. In 2018, 893 million people in 49 countries required preventive chemotherapy to stop the spread of the infections. Over 7.7 billion doses have been distributed since 2000. Since that time, 597 million people no longer need preventive or chemotherapy. Preventive measures include chemotherapy, permethrin treated bed nets, vector control, indoor spraying, air conditioning, and a combination of three medications in a cocktail that eliminates the infection within a few short weeks.

CHAPTER 10: MALARIA

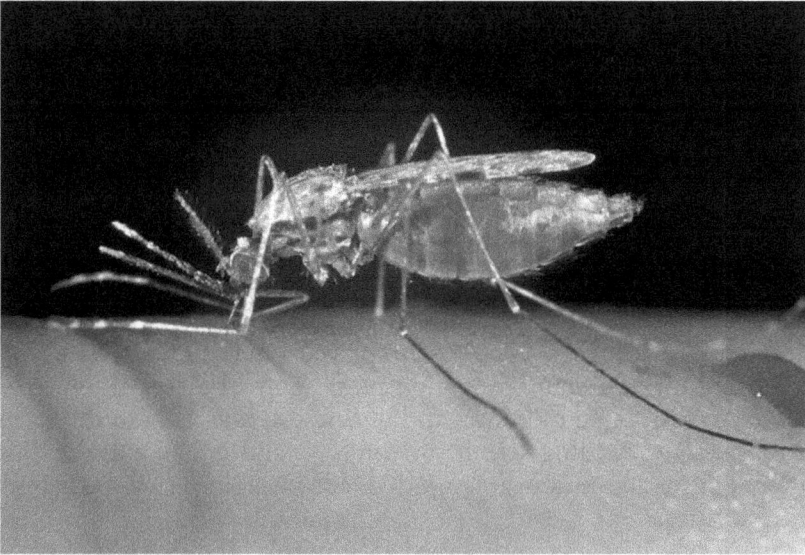

Photo: Female *Anopheles gambia* discharging excess fluid and heat while blood feeding. Photo by Jim Gathany – CDC/AFPMB

WHEN I first began to research this portion of the book, I was reminded that the previous amount of work, research, and monies spent by NGOs and governments was staggering. I decided to maintain the theme of this book as a 'primer' to avoid lengthy chapters from the seemingly unlimited volumes of information about malaria. Therefore, I have selected what I feel will provide the reader with some of the more important points regarding the biology of the mosquito and the protozoan responsible for the single most deadly disease humans have ever faced.

Malaria, from the Italian *mala aria* or 'bad air' is a leading cause of human death and illness transmitted by mosquitoes. The protozoan parasite *Plasmodium* is transmitted through the blood-feeding of the female *Anopheles* mosquito. There are five species within the genus *Plasmodium* responsible for human infection. *Plasmodium falciparum* and *Plasmodium vivax* pose the greatest threat of severe malaria infections to humans. These two parasites account for more than 95% of all malarial infections. *Plasmodium falciparum* is found in sub-Sahara Africa while *P. vivax* is not as common in sub-Sahara Africa but is endemic in parts of Asia, Oceana, and Central and South America. *Plasmodium falciparum* is by far the deadliest of the five species. The other three are *Plasmodium malariae, Plasmodium ovale*, and *Plasmodium knowlesi. Plasmodium vivax* and *P. ovale* may persist in the human liver for years and emerge as a form of relapsing malaria. *P. malariae*, if left untreated, is known to persist in the human body for decades. *P. knowlesi* is limited to southeast Asia typically found in macaque monkeys and other forest primates and has been linked to human cases although limited in infectious rates.

The discovery of the etiology of malaria followed decades long research culminating in the recognition that malaria was indeed transmitted by mosquitoes. Malaria became endemic in the United States following the return of Civil War soldiers who served in a southern post and spread the disease to more northerly states upon returning from the war. Malaria finally began to disappear during the 1940s where a combined effort utilizing anti-mosquito strategies, improved medical care, and the draining of swamps and wetlands. Technology also played a major role in reducing the cycle of mosquito-human interaction in the United States. The increased use of air conditioning of homes and businesses, especially in the warmer, subtropical states, adding screens to windows and doors, and the change in lifestyle from an outdoor to an indoor life brought on by the invention of the television.

Species of *Anopheles* once served as important transmitters of malaria. These include *An. quadrimaculatus* in the eastern U.S., *An. freeborni* along the western coastal U.S., and *An. punctipennis* along the Rio Grande of our southernmost boundary into Mexico. A fourth species, *An. hermsi*, perhaps not as well-known, was a vector in California. It has been estimated that approximately 800 to 1200 cases of malaria are introduced into the United States each year. Globally, a few of the most important species are *An. gambia* and *An. funestus* in Africa, and *An. darlingi* in South America. *An. gambia* stands out as the most dangerous of all the species because it has been responsible for much of the malaria cases and deaths in Africa.

These parasites are obligate and develop in both vertebrates and mosquitoes of the genus *Anopheles*. There are approximately 430 mosquito species within the genus where roughly 40 have been identified as transmitters of malaria. Recent genetic studies place the divergence of these parasites as much as 40,000 to 60,000 years ago with the final divergent strains occurring somewhere around 4,000 to 6,000 years ago. The *Lavernea* family of protists known to exist within African apes is the source of the evolved parasites having made the jump from primates like gorillas and chimpanzees to humans.

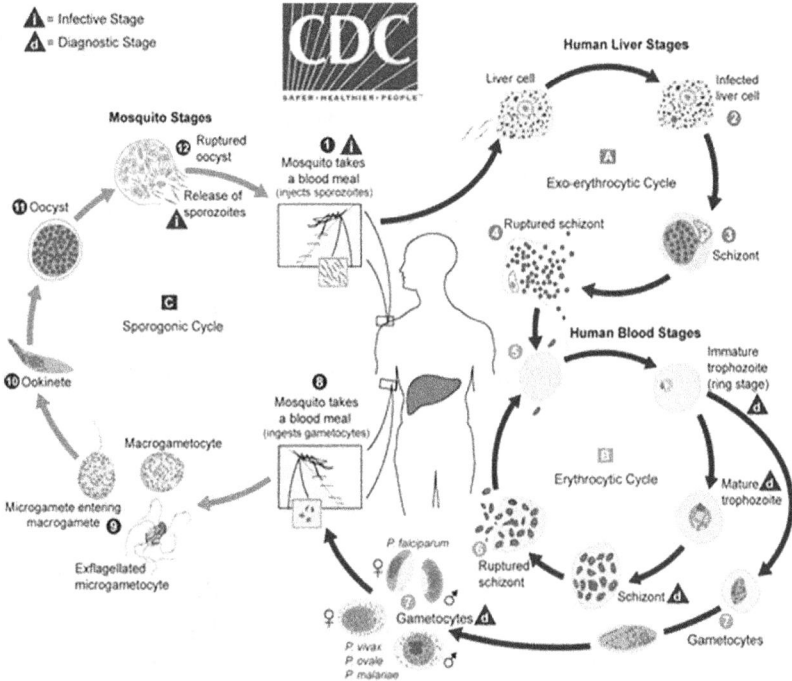

The parasitic lifecycle shown here involves three distinctive stages.

The exo-erthrocytic cycle or human liver cycle, the erthrocytic cycle or the human blood cycle, and the sporogonic cycle, the infectious stage in mosquitoes. When a human is infected by the sporozoites they migrate to the liver and lodge within the cells. Infected liver cells reproduce shizonts that multiple and eventually emerge from the lysed liver cells and enter the red blood cells. Further maturation results in the production of gametocytes which give rise to male and female cells. It is these gametocytes that the mosquito ingests when feeding upon an infected person. Maturation reaches the final infectious stage where sporozoites migrate to the salivary glands and are ready to be passed on when the mosquito takes its next blood meal.

In 2018, according to WHO, *P. falciparum* accounted for 99.7% of malaria cases in the African region, 50% of cases occurred in the southeast region, 71% of cases in the eastern Mediterranean, and 65% in the western Pacific. In the Americas, WHO reports that *P. vivax* is the predominant parasite causing 75% of malaria cases. Malaria in the western Hemisphere was likely introduced during the periods of exploration and colonization primarily from Europe, the slave trade, and the expanse of international trade.

According to the world malaria report released in December of 2019, 228 million cases of malaria were identified with a similar number of cases in 2018 of 231 million. Estimated number of deaths was 405,000 in 2018 and 416,000 deaths in 2017. The regions of Africa carry a disproportionate share of global malaria where 93% of malaria cases and 94% of malaria deaths from 2018 estimates occurred in WHO African regions. Children under five years of age are the most vulnerable age affected by malaria. In 2018, 67% or 272,000 of all malaria deaths occurred in this age group.

Approximately half of the world's population is at risk of malaria infection. Within these groups are infants and children, pregnant women, people with HIV/AIDS, non-immune migrants, and travelers. In climates where the mosquito lifespan is long, the parasite has ample time to complete its development in the mosquito. In addition, areas where the mosquito tends to select and feed upon humans rather than other vertebrates leads to increased infectivity. These conditions are reflected in the fact that 90% of the world's malaria cases are in Africa. Transmission is dependent upon climatic conditions that may affect the quantity of mosquitoes and their ability to survive. Primarily, rainfall patterns, optimal temperatures, and humidity levels favor the mosquito therefore the developing parasite in those sub-Sahara environments. Malaria can result in levels of illness from mild to severe and death can be a result. Much of the levels of severity depends upon the species of *Plasmodium* responsible for the infection.

Preventive measures such as vector control using insecticide-treated bed nets and indoor spraying has reduced the contact of humans and mosquitoes and has lowered the global number of cases and deaths nearly in half. Young children who go to bed uncovered by bed nets were exposed to repeated bites by *Anopheline* mosquitoes. The preventive treatments have cut into the exposure and therefore the infectious rate of young children. By 2018 approximately half of all the people were protected from insecticide treated bed nets as compared to 29% in 2010. However, since 2016 the effectiveness of bed nets has not changed. Unfortunately, there has been insecticide resistance to at least one of the four commonly applied insecticide classes as reported by 73 countries between 2010-2018. Resistance to anti-malarial drugs has been an additional recurring problem. The efficacy of therapeutic drugs used in the 1950s and 1960s have been reversed undermining progress in child survival. A positive note lies in the reports of one vaccine with an efficacy of 40% in children. The vaccine, (RTS,S) is the first vaccine that has significantly reduced malaria infections by acting against *P. falciparum*. Children in selected areas of Ghana, Kenya, and Malawi were given the first of four injections starting in 2019. The injections are spaced apart every 6 months starting at the age of 6 months. The vaccine was first reported by its maker GlacoSmithKline (GSK) in 1987 along with researchers from Walter Reed. In 2001 GSK and PATH with support from the Bill and Melinda Gates Foundation developed the vaccine to target infants and children. In 2014, the efficacy was reported to have a positive effect upon 40% of the recipients. The downside is that the vaccine wanes over time. Phase 4 is now being tested in the above-mentioned countries. The vaccine interrupts the sporogonic cycle by interfering with the parasites (sporozoites) ability to enter the liver during the exo-erthrocytic cycle.

The virulence of malaria does contain a silver lining that dates back thousands of years. Those Africans who are heterozygote for sickle cell, a common condition amongst Africans, provide some protection from *P. falciparum*. However, these individuals are carriers and may pass on the gene for sickle cell. Those unfortunate to be homozygote for the disease suffers from numerous health related conditions and often have shortened lifespans. Native people in Africa may develop a partial immunity from years of exposure if they survive their early childhood. It does not provide complete immunity but does reduce the risk of a severe infection.

CHAPTER 11: VIRUSES

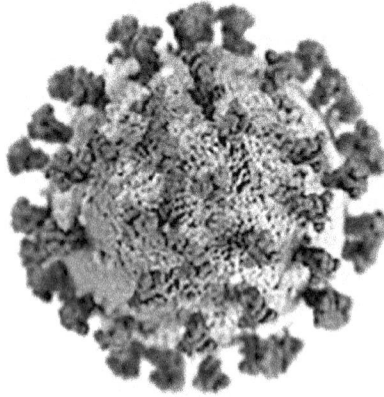

Photo: Corona Virus provided by the
Center for Disease Control and prevention (CDC)

VIROLOGY is a relatively new scientific field of study when compared to the field of genetics. Knowledge obtained through research has improved are understanding of viruses from which science continues to build upon. However, viruses remain as a controversial issue in that it is unclear as to whether to classify these organisms as living organisms when placed up against most any criteria that defines life. Interestingly, the role viruses played in driving evolution as a precursor of life on earth has some merit. Consequently there remains no consensus in determining the origin and role of viruses. It has been hypothesized that viruses started out as a free-living organism that later adapted to a parasitic lifestyle. It is unclear that viruses may have served as a precursor to the evolution of life on earth. If they played a role was it tangential to the dominance of bacteria billions of years ago or was it the adapted behavior of the virus responsible for the evolution of life, or both. Genetic studies also suggest that viruses adapted to life by invading between the cells of a host or within those cells. This may have lent itself to its parasitic role in humans and most all forms of life.

What we do know is that viruses replicate only within a host because they do not obtain the mechanism to replicate or translate proteins on their own and therefore rely upon the ingredients and cellular mechanisms of their host. Viruses do not generate the energy bound ATP molecule necessary for life, a process found in living cells. Since viruses do not have the capability of translation, no proteins can be synthesized. Viruses are said to be about 95% protein and 5% genome. The protein structures of viruses, the capsid and spike formations, rely upon their ability to enter a cell and take over that cell for the purpose of reproduction and the translation of proteins. Failure to manufacture the necessary proteins in the host cell eliminates that virus from its host. One virus particle that enters a cell can generate thousands of virions, lyse that cell, and travel to adjacent cells to continue their reproductive process.

Viruses are a small particle of less than 200 nanometers (nm). One nm is equivalent to one billionth of a meter to give you some perspective as to size (pointed out in Chapter one). The genomes are either single or double stranded nucleic acids of RNA or DNA. Over time, the genome of viruses has gone through a series of reductions and has occupied the human genome. HERVs, or human endogenous retroviruses (RNA genomes) occupy as much as 8% of the human genome, although inactive, they can be reactivated and are suggested to be associated with human illnesses.

Let me be clear, the information regarding mosquitoes and viruses, the arboviruses, behave differently than many of the viruses that infect humans that are not associated with mosquitoes. One case in point, the Sars-CoV-2 virus we are all keenly aware of appears to have jumped from bats to possibly an intermediate vertebrate host and then adapted to infect humans and other mammals. Pets have tested positive (dogs), the commercial enterprise surrounding mink farming has been invaded, and captive gorillas at the San Diego Zoo have also been infected. This respiratory illness does not follow the transmission spread associated with arbovirus infections. In the case of Covid-19, a virulent virus, is spread from human to human, and does not appear to require a reservoir host to continue the spread although a host or hosts may exist. However, recent studies indicate that mink and humans can spread the virus back and forth. In the case of mosquito transmitted viruses, the mosquito remains infected for life, albeit a relatively short lifespan, while the human recipient, once clear of the virus develops an immunity for life. This result has been documented for cases involving Zika, dengue, chikungunya, West Nile virus and so many other arboviruses. An arbovirus requires a host, the mosquito, and a vertebrate dead-end host allowing this cycle to repeat itself from mosquito to human back to mosquitoes and again transmitted to humans.

Mosquitoes remain the leading transmitter responsible for many human infections globally, while ticks have become the most zoonotic infections throughout North America. A zoonotic disease is an infection caused by a pathogen that has made the jump from a non-human animal, usually a vertebrate animal, to humans (Covid-19). Anthropomorphic behavior is responsible, at least in part, for the increases in pathogen- vector- host diversity. We can witness these ecological changes manifested through climatic alterations, the loss of habitat, and decreased biodiversity. These changes and many more are having a tremendous impact upon the ecosystem. Neglected tropical diseases are a result of some of these ecological changes characterized by those people living under some of the poorest conditions. They are affected by tropical deforestation practices, fractured forest habitats, urbanization, questionable agricultural practices, and population growth and migration. These ecological displays have led to an increase in human-pathogen contact especially when novel microorganisms emerge. It should be noted that not all pathogens or their carriers are capable of transmission of disease. Genetic make-up influences the interactions of pathogens, vectors selected, and sources of reservoirs incorporated to provide successful transmission of vector-borne illnesses. The complexity of ecological factors test's sciences ability to prevent and control these pathogens and requires a change in strategies over time.

Biological transmission from insects, fleas, ticks, flies, and mosquitoes, requires the vector to imbibe on a blood meal from a previously infected host. It must then continue to develop and replicate, pass through barriers including entering the salivary glands of the mosquito, and then reintroduce the virus into a susceptible animal through blood-feeding. Those mosquitoes that accomplish these stages is said to have a vectorial capacity. Laboratory tests may reveal that a larger number of mosquitoes showing this capacity to transmit does not equate with those mosquito species in the wild.

Since it is my intention and hope that emerging biology students will expand their knowledge of mosquitoes as vectors, it seems proper at this point to define a few terms for their use later in their studies. To begin with, zoonotic disease where the virus can make the transition from an animal source to humans and back and forth repetitively is a descriptor used for defining arboviruses. Covid-19 is a zoonotic disease, where the infectious agent made the jump from bats, and quite possibly to pangolins, and then to humans. This hypothesis considers the contact between humans and bats, and humans and pangolins, a more likely source of the zoonosis. The intermediary host, the pangolin, appears competent enough to generate sufficient virions prior to being transmissible to humans. It is too early, at the time of this writing to be definitive regarding the zoonotic episode that has become pandemic. The term epizootic references a disease that is present in animal populations striking a balance between the host and the virus. In humans it is often characterized as endemic to a population, geographic area, race, or other limitations. Outbreaks do occur, they are often short lived-in duration, if it becomes widespread, an epidemic, or epizootic event, of greater proportions are referred to as a pandemic. West Nile virus (WNV) is endemic and epizootic throughout the United States. The last major geographic area where an outbreak of WNV occurred was in Dallas, Texas in 2012. Malaria is endemic and an epidemic throughout Africa arising in several countries, especially during the rainy season. These terms are at times interchangeable and sometimes confusing.

Transmission refers to the passing of the infectious agent from one host to another. Here we are addressing the modes of transmission that takes place amongst arboviruses. In the mosquito, transmission occurs through horizontal, vertical, or venereal pathways.

Horizontal transmission occurs when the virus passes between two in-dividuals that are not related as parents or offspring. This transference can occur between the same species of mosquito or different species. West Nile virus is found in over 60 species of mosquitoes. Only a few can develop a vector capacity to deliver a sufficient viral load to humans. The main genus is the *Culex* mosquitoes, although other genera have been tested for the virus but seem incapable as a transmitting host. Primarily it is the northern and southern house mosquitoes, *Culex pipiens* and *Culex quintquefaciatus* that are the prime hosts and vectors of WNV.

Vertical transmission involves the infected host, the female mosqui-to, to directly transfer the virus throughout all stages of development to her progeny regardless of the mechanism adopted in the transference. This includes male progeny and is not limited to the female progeny that later become blood-feeding females.

Venereal transmission is a form of horizontal transmission where the infected male from the vertically transferred virus is passed on to females during copulation. To my knowledge, I am not aware of any research that has shown this process to occur in the wild. However, it stands to reason that this is a viable form of transmission amongst mosquitoes.

Hematophagus arthropods, principally the female mosquitoes, are a major cause of disease. Their ability to become vector competent requires several steps in the process. Blood-feeding requires the uptake of sufficient virions that pass through, survive barriers, and continue development prior to reaching the salivary glands where the pathogen is then passed to human hosts. The resultant viral illness having been effectively passed along is of-ten characterized as either hemorrhagic or encephalitic.

Chapter 12: Mosquito transmitted Diseases

NOW that you have increased your knowledge about mosquitoes, the role they play as vectors, the accompanying pathogens, and hosts, let us look at some of the leading mosquito hosts that also serve as transmitters of viruses and the diseases that are passed on to humans.

There are over 130 arboviruses transmitted by mosquitoes. In addition, infections have occurred from blood transfusions, organ transplants, perinatal transmission, breast feeding, venereal transmission, and laboratory exposure to a pathogen. Three of the five arbovirus genera are responsible for most of the diseases associated with mosquito bourn illnesses. *Flavivirus*, is responsible for Zika, yellow fever, dengue, Japanese encephalitis, and West Nile fever. *Alphavirus* is the virus that is responsible for chikungunya and Eastern equine encephalitis and Western equine encephalitis, and *Orthobunyavirus*, related to Jamestown Canyon virus and LaCrosse virus. It is of interest to note that in most of these viral diseases those infected are predominately asymptomatic.

Previously alluded to in an earlier chapter, there are three main mosquito genera associated with the many illnesses transmitted by mosquitoes to humans. The genus *Aedes*, is associated with the global transmission of dengue fever, chikungunya, lymphatic filariasis, Rift Valley fever, yellow fever, and Zika. The genus *Anopheles*, the deadliest of pathogen, are transmitters of malaria and lymphatic filariasis. The third genus, *Culex*, transmits diseases like Japanese encephalitis, lymphatic filariasis, and West Nile virus.

Species like *Aedes aegypti* and *Ae. albopictus* are transmitters of dengue, yellow fever, chikungunya, and Zika. *Anopheles gambiae* and *An. funestus* in Africa, *An. darlingi* in Central and South America are transmitters of malaria. *Culex pipiens* and its associated complex of species is the main carrier of West Nile virus.

It should be clear to the reader that mosquito transmitted diseases are on the rise and have the potential to cause serious epidemics. We continue to garner information regarding the ecological changes that often facilitate transmission of pathogens between the mosquito world and the human population. With an ever-changing environment, there are indications that novel, primarily unheard-of viruses transmitted by mosquitoes could very well be the next outbreak of disease. Several of these arboviruses are currently restricted to small regions of the world and have affected very few humans. The primary geographic locations of these restricted viruses are in South and Central America, and the sub-Sahara of Africa. However, not far from home, the Florida Everglades is home to the Everglades virus. It is the *Culex cedecei* mosquito that is suspect in the transmission within the region. The Everglade virus is an encephalitic illness resulting in fever, headache, and myalgia.

Another of these viruses is the Mayaro virus suspected to be transmitted, not by one of the beforehand mentioned 'big three' genera but rather a species from the genus *Haemagogus*. It is difficult to label some of these viruses as novel since the Mayaro virus (*Togaviridae: Alphavirus*) was first isolated in 1954, several decades ago on the island of Trinidad. Symptoms of this infection seem to parallel that of chikungunya (*Togaviridae: Alphavirus*) including fever, rash, myalgia, and arthralgia. There have been outbreaks in Venezuela in 2010 involving 77 cases and one case in Haiti in 2015. Were these cases infected from locally transmitted mosquitoes? Haiti does not have a population of non-human primates to serve as a reservoir or jumping off point from an earlier blood-feeding episode. The same individual was also co-infected with dengue, probably from *Ae. aegypti*. Therefore the source of the transmission is unknown. Finally, to add a bit of confusion as to the source in Haiti, a migrating bird was found to harbor the Mayaro virus having migrated from Colorado. There are several *Alphaviruses* that are endemic to the Americas and the potential exists for recombinants forming a new strain although virologists believe this would be an unlikely scenario.

Viruses known to exist in local transmission cycles have since been distributed worldwide. Mosquitoes that are capable of a high vectorial capacity, such as *Ae. aegypti* and *Ae. albopictus*, coupled with increased incidence of human travel, increased viral-mosquito interaction, and with human-mosquito contact rates increasing, contribute to outbreaks throughout the world. Sound strategies to combat invasive mosquitoes through increased surveillance methods, improve vector controlling methods to mitigate outbreaks, and affordable cost control of diagnostic testing, effective therapeutics, and vaccines should be on everyone's agenda, otherwise, the pattern we have been monitoring will only continue to get worse.

CHAPTER 13: WEST NILE VIRUS

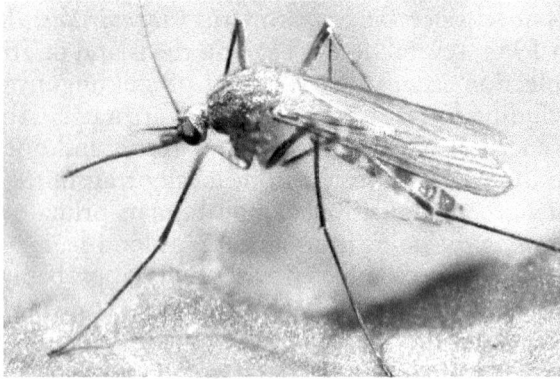

Photo: *Culex pipien*, the northern house mosquito.
One of the most ubiquitous mosquitoes and a major contributor to the
vectoring of West Nile Virus. Photo by Ary Farajollahi, Bugworld.org.

WEST Nile virus was first isolated in 1937 in Uganda and spread into
Europe, the Middle East, and parts of Asia. Outbreaks were sporadic and
it was not until 1996, in Europe, that larger outbreaks began to occur.
The Western Hemisphere was spared from this illness until an outbreak
occurred in August of 1999 in the borough of Queens in New York City.
The New York City Department of Health was made aware of an unusual
number of meningoencephalitis cases and later determined that the cause
was probably from arbovirus transmission. Earlier that summer, in July,
an unusual number of dead American crows (*Corvus brachyrhynchus*) from
an avian infection was observed to go along with reports from New York
State, Connecticut, Maryland, and New Jersey. Crows, jays and 14 other
species of wild native birds resulted in an epornithic event (an infectious
outbreak in wild birds). In the end, 62 human cases were identified that
led to 7 deaths. Those infected ranged from 5 to 90 years old. This was the
first time that WNV was introduced into the Western Hemisphere. The
sporadic outbreak appeared to cease when no new cases were reported to
the health authorities after September 22, 1999.

Today we can report that WNV is enzootic in birds and several species of mosquitoes (over 60 in the U.S.) that have tested positive for the
presence of the virus. Only a handful of the 60 plus mosquitoes that have
tested positive for the virus display the capability of accomplishing vector
competence and therefore reaching a vectorial capacity necessary to transmit an infectious viral load to the host. Vector competence is held in check
from the failure of the virus to breach internal barriers within the mosquito
as well as an antiviral response to the viremia attained from feeding on viral

infected blood. Once viral competence is checked, then vectorial capacity, reproductive ability within the mosquito cannot drive the emergence of infection in its blood fed upon host. Two notable exceptions, that are responsible for outbreaks around the country are species of mosquitoes from the *Culex* complex, being *Culex pipien* and *Culex quinquefasciatus*. The expansion westward in the United States is likely from the dispersal of infected birds and the role played by *Culex tarsalis* as an amplifying host of West Nile fever that had a great deal to do with the spread. The transmission cycle involves birds, mosquitoes, horses, and humans. In a repetitious fashion, re-infection from birds to mosquitoes and back to birds continues, where the birds serve as reservoirs and the primary amplifying host of the virus. The mosquito serves as a bridge vector to horses and humans' where they, in turn, serve as dead-end hosts. A dead-end host is generally characterized as a recipient that does not produce an infectious viremia.

The speed of spread of this virus is more than likely due to several factors. Primarily, the reservoir hosts, birds, can carry the virus over great distances. This includes the many migratory birds that travel thousands of miles. Adding to the cause and effect is the warming climate expanding the range of the avian populations introducing the virus to compatible mosquito species. The mosquito *Culex pipiens* complex are a ubiquitous species and therefore are available as a bridging host once the infected birds enter a virus-free habitat and are fed upon by the mosquitoes to begin the cycle of transmission.

Monitoring the spread of WNV required surveillance as to which states report the infection, and what birds tested positive in that state, along with mosquito testing to develop vector control strategies. The speed of spread in the United States has been documented by state health authorities. In 2000, there were 2 deaths, 12 states that reported infected birds, and 5 states that reported infected mosquitoes. In 2001, 10 states reported illnesses. By 2002, 39 states reported illnesses and some 4756 cases. It did not take long for the virus to be transmitted to 48 states and 7 Canadian Provinces, Mexico, Caribbean Islands, and Columbia. All of this was reported in 2004. In 2000, unusual numbers of bird deaths occurred along the eastern seaboard from New Hampshire to North Carolina. In 2001, 12 states and the District of Columbia reported bird fatalities. In 2002, 14,172 bird fatalities were reported where 7719 were American crows and 4948 were blue jays (*Cyanocitta cristata*). Ninety-two other species of birds accounted for the remaining 1455 fatalities. The rapid spread was not confined to birds and mosquitoes, in 1999, 31 horses and 4 deaths occurred. In 2000, 58 horses primarily in New York and New Jersey were infected. In 2001, 733 equine cases were reported where 66% occurred in Florida. In 2002, 9144 horses were reported ill or dead from just 6 states spread out over Illinois, Texas, Mississippi, Indiana, Kansas, and South Dakota. In 2003, an equine vaccine was introduced, and the number of infections dropped dramatically. Today there are several equine vaccines for horses while no vaccine has been given approval for human use. Human vaccine development is far more troublesome to ensure a safe and effective vaccine. Taken from a scientific evaluation, we are dealing with a ssRNA virus with at least 5 lineages and multiple mutations. There is the added concern regarding co-occurrence with other viruses from the same family of viruses, the *Flaviviridae*. The same family that is responsible for dengue, yellow fever, Zika, Murry Valley encephalitis, St. Louis encephalitis, and Japanese encephalitis. From a cost-benefit or economic impact approach, 80% of WNV victims are asymptomatic while most of the remaining cases are mildly symptomatic showing up in one of every five cases. Symptoms include a low-grade fever or headache. It is within the older generation of people or those with immune-compromised conditions that are most vulnerable. Severe cases occur in one out of one hundred and fifty cases. Those individuals may have an advanced fever, encephalitis, or meningitis.

There have been about 2300 fatalities and thousands of hospitalizations from WNV infections since 1999 to go along with a decline in avian populations. During my short stay in Hawaii in the fall of 2018, it was pointed out to me that on several of the islands I had visited that the number of birds, both in abundance and richness, was being drastically reduced from WNV and avian malaria. For the first eleven months of 2020, 451 cases have been reported in the United States along with 27 deaths. Blood donors have tested positive in 112 people. Human infections have been reported in all but four states. Six other states reported non-human infections through veterinary cases involving mosquitoes, birds, and sentinel animals. The latest information regarding positive testing in wild and domestic birds has climbed to over 300 species.

Chapter 14: Dengue

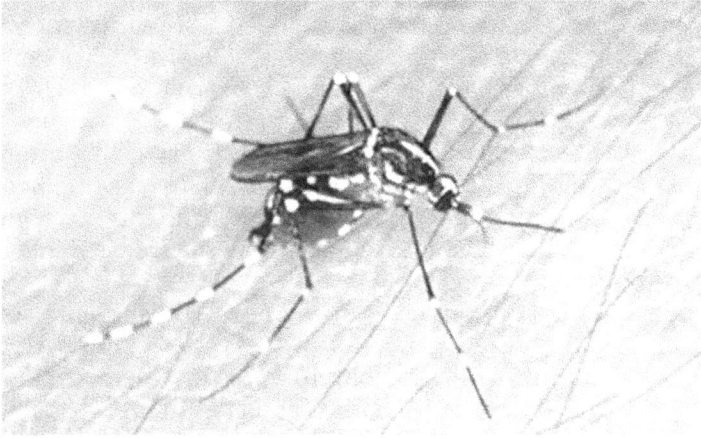

Photo: Female *Aedes albopictus* an invasive mosquito probably transported through the importation of used tires. Note the outstanding markings that are used to identify this species. The dorsal view of the thorax and head display a pale single stripe.

ALTHOUGH dengue fever is not the leading cause of death when compared to the deaths from malaria approximately 50% of the global population is at risk of being infected through mosquito transmission of the virus. This occurs primarily in tropical and subtropical regions of the world where most people have a greater chance of being infected.

The earliest record of a dengue-like illness was recorded in a Chinese medical report from the Jin Dynasty (265-420 AD). Writings also suggested that it may be linked to flying insects. Symptoms included myalgia, muscle pain that was not caused by a non-disease illness, and arthralgia a joint pain where bones are connected, again, not caused by an underlying illness. The first reports of dengue fever epidemics were described in Asia, Africa, and North America in the 1780's. Identification of the disease was described based upon symptoms that led to the naming of this disease in 1779. The related term, 'breakbone fever', appropriately described by Benjamin Rush was based upon earlier descriptions of myalgia and arthralgia symptoms. It was not until the 20th century that the etiology, the causes and origin of disease, and the mode of transmission were determined.

Amplification of the disease has several components as to the reasons for such a large expansion throughout the world. The leading cause is correlated to poor living conditions and the ever increasing over population in regions of the world. When these conditions are coupled with poor vector control, polluted and stagnant water allowing for mosquito breeding sites, climate change, increased transmission through mosquito behavior, and a staggering increase in travel to endemic areas has had a major impact upon the spread of this disease.

Dengue virus (DENV) is a single stranded RNA virus (ssRNA) as are most of the arboviruses. This includes West Nile virus, Zika, and yellow fever. They are members of the genus *Flavivirus* all of which cause encephalitis in humans. There are 4 strains or serotypes of dengue. They are denoted as DENV1, DENV2, DENV3, and DENV4. The more serious and life-threatening infections lead to dengue hemorrhagic fever (DHF) and dengue shock syndrome (DSS). In 2013, a fifth virus was reported involving a sylvatic cycle where DENV1 – DENV4 have been incorporated into a mosquito/human cycle.

There are several reasons as to how a novel virus might evolve. The most obvious would be the potential for genetic recombination of genomes. The greater the circulation of serotypes with increased human behavior will likely lead to genetic changes and a resultant broader diversity of viruses. To go along with potential recombinants, we must also consider the mechanism of natural selection and viral genetic bottlenecks, as factors that lead to a novel serotype of the virus. Dengue viruses as ssRNA will be more likely to have a higher mutation rate than DNA viruses. The rate of mutation and the accumulation of these mutations may also lead to a new serotype of dengue. It is worth noting that thousands of years ago, the dengue virus evolved in non-human primates and later made the move to humans. This sylvatic cycle still exists today and is maintained in the non-human primates as reservoirs along with the *Aedes* mosquitoes. The precise reason for the emerging DENV5 serotype is unclear. It is probable that the sylvatic form of DENV5 has been in circulation among the non-human primates for centuries prior to its jumping to humans. Phylogenetic studies have placed DENV5 in a genetically similar pattern as determined for the other four serotypes suggesting a common ancestral link.

The first vaccine came to market in 2015 and is used to target the endemic regions of the world. Individuals from ages 9 to 45 were the predominant recipients of the initial dose. Other vaccines are in various phases of development. The main problems lie in the preferred development of a tetravalent vaccine, the difficulty in DENV2 resistance, and the phenomenon known as antibody dependent enhancement (ADE). The first infection often results in a mild or even asymptomatic response in humans. A major problem lies in the acquiring of a second infection. If you are infected by one of the four serotypes of dengue, you build an immunity to that one serotype. It remains that the infected individual is still susceptible to the other three viruses, each of which may exacerbate a more severe reaction than the initial infection. It is this phenomenon response that the currently available vaccine should not be given until after the initial infection was attained otherwise a far more serious form of dengue (DHF or DSS) is likely to occur. This process requires a rapid diagnostic test prior to receiving the vaccine that would follow the initial infection.

There are no specific treatments in the form of medications nor are there sufficient early detection tests to identify the more severe forms of dengue fever. The current control of disease transmission follows strategies established for vector controlling of mosquito populations. Currently, 70% of the illness lies in Asia. In 2000, 505,430 cases were reported. In 2010 the number rose to 2.4 million cases. The latest data shows 4.2 million cases in 2019. Deaths related to dengue went from 860 to 4032 during the same period and is likely to be an underestimated figure which is true with most of the arboviruses transmitted to humans.

The primary vector is the mosquito *Aedes aegypti*, the very same mosquito associated with the Zika outbreak, chikungunya, yellow fever, and other arbovirus transmissions. *Ae. aegypti* is a daytime feeder usually seeking its blood meal at dusk and at dawn. This is an urban insect known to be in and around concentrations of human dwellings. There are other mosquito species that serve as a secondary transmitter of dengue. The species of most concern following *Ae. aegypti* is *Ae. albopictus*. Unlike *Ae. aegypti* known to thrive in the warmest of climates, *Ae. albopictus* is a forest mosquito that has a cold tolerance existence when comparing the northern most range of the two species. *Ae. aegypti* has been found in the sewer systems during cold months in Washington D.C. while *Ae. albopictus* has established populations in the colder southernmost counties of New York. *Ae. albopictus* has been found in 32 states where their cold tolerance as an egg or adult has led to established populations and an ever-increasing range.

Once the mosquito takes a blood meal, the virus moves into the midgut where the virus replicates and eventually lodges in their salivary glands. The incubation period is anywhere from 8 to12 days with an optimal temperature range of 25 to 28 degrees Celsius. When the temperature range becomes too high or too low the reproductive process is slowed. Other factors in the development and eventual transmission of the virus deals with viral genotype and the viral load. Once the virus reaches an infectious state, the female and progeny, through vertical transmission remains infective for life.

The CDC reports that there are over 400 million infections each year where approximately 25% become ill and 22,000 die. Compounding the problem is the antigenetic differences between the four serotypes which may cause up to four different infections in a single individual. Typically, symptoms during the infection period involve a variety of aches and pains, eye pain, muscle joint pain and bone pain. Included in symptomatic patients are nausea, vomiting, and a rash. These symptoms typically last anywhere from 2 to 7 days. Following any fever are stomach pain and tenderness, vomiting at least three times in 24 hours, nose and gum bleeding, blood in vomit or stools, tiredness, restlessness, and irritability. There have been reports of vertical transmission from mother to child during pregnancy. There has only been one documented case of transmission from mother to child from breast milk. In 2020, two hundred and fifty cases have been documented in the United States where 181 are from travelers returning from endemic areas. Sixty-nine cases have been designated as infection from a locally mosquito-human encounter. All 69 cases are reported from Florida while 436 cases occurred in Puerto Rico.

As we go through the various illnesses, it should become clearer to the reader that there exist many similarities between viral disease transmitted through the blood feeding activity of the mosquito and the knowledge we have accrued in the last ten months of 2020 regarding the Covid-19 virus.

CHAPTER 15: ZIKA

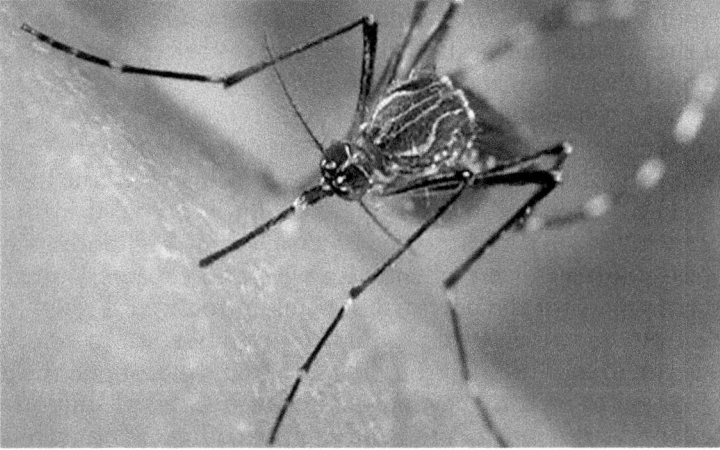

Photo: *Aedes aegypti* known as the yellow fever mosquito. In the 21st century it is often referred to as the Zika mosquito. Note the identifying markings on the scutum that distinguishes this species from *Ae. albopictus* shown below.

THE Zika virus was first identified in 1947 from a sentinel rhesus monkey in the Zika Forest of Uganda while scientists were conducting surveillance efforts targeting the presence or absence of the yellow fever virus. In 1948, the mosquito, *Aedes africanus*, captured from a tree platform in the Zika Forest was found to harbor the Zika virus. The first human infection was recorded in 1952 in Uganda and later in the United Republic of Tanzania. The viral disease was treated as a neglected tropical illness and little attention was given to this novel disease. It was not until 2007 when a major outbreak of the illness occurred on YAP Island in the Pacific that had suddenly grabbed the attention of the medical world. It was estimated that 75% of the 270,000 YAP residents had been infected. This outbreak garnered the attention of health agencies and concern began to grow over another virus that made the jump from a sylvatic cycle to human infection that had previously been reported and recognized decades earlier. In 2013, the island of French Polynesia and four other island nations reported outbreaks of the Zika illness adding to the growing concern. In 2015 an unexpected and large outbreak of the Zika illness occurred in Brazil and soon after was reported to have been identified in neighboring countries throughout South and Central America with a risk to the United States. The concern grew over the potential of being introduced through travelers and trade from known endemic countries into the states. In Brazil, health agencies began to monitor the troubling secondary affect upon the unborn. Numbers began to increase beyond the recorded mean of stillborn births, fetal losses, premature births, and neurological complication including encephalitic paralysis illness of Guillain-Barre' Syndrome (GBS), microcephaly, neuropathy, and myelitis.

In Brazil, between 2015 and 2019, one and a half million cases of Zika were identified along with a 19% increase in GBS. To date, only non-human primates and humans are infected through the transmission of the virus from mosquitoes. The primary mosquitoes in South America involve both *Aedes aegypti* and *Aedes albopictus*. In the Pacific Islands, the virus has been found in two additional *Aedes* mosquitoes, *Ae. polynesiensis* and *Ae. hensilli*. The genus *Aedes* are known to be daytime feeders peaking during dusk and dawn hours, a most vulnerable time for children. They are also associated with natural or artificial container breeders that are found near populated areas thus increasing the contact ratio between the mosquito and humans. Research has determined that there are three lineages, two from Africa and one from Asia. The Asian strain appears to be the cause of the viral fever in the Americas as determined through pathogenesis studies.

Difficulty in determining the virus infecting humans without sound antibody testing lies in the fact that the viruses often share similar or same non-specific symptoms. In both dengue and Zika, most are asymptomatic and therefore the actual number of reported cases is greatly underestimated. Zika symptoms include fever, rash, joint pain, red eyes, muscle pain, blood in the urine, along with the concern over the neurological disorders in some adults and certainly displayed in the unborn and neonatal.

The mosquito becomes infected once it takes a blood meal from an infected human. It has been suggested that the first 7 days of human infection is the greatest period of infectivity for the feeding mosquito. Once that occurs and the virus passes through the mid-gut and salivary barriers of the mosquito it becomes infectious and is likely to pass the virus on to an uninfected human. Here we have a repetitive horizontal transmission between

two animals that meet the required idea of an arbovirus. There are forms
of vertical transmission between mother and fetus. This transmission is the
likely mode of the neurological illnesses GBS and microcephaly. Other
brain defects have been noted as well. There is no confirmation that breast
milk is a mode of transmission even though the virus has been found in
the infected mothers' milk. Venereal transmission, a form of vertical transmission, has been documented and is likely passed on to a partner from
an asymptomatic individual. It has been determined that the Zika virus
lasts longer in semen than in vaginal, urine, or blood fluids. The number
of cases determined through venereal transmission is quite low. As is the
case for many arboviruses, the human infection often results in a lifetime
immunity while the mosquito remains infected for its entire life. A positive
test for the presence of the virus in body fluid has been isolated from tears.
Little information is available as to the degree of transmissibility of this
mode between humans.

In addition to mosquito abatement strategies in known epidemic sites,
the development of a vaccine is probably the best means of controlling the
Zika virus. The problems associated with such a development has brought
science to the point where, according to WHO, some 18 different vaccines
are in pre-clinical or early phases of clinical trials. The difficulty is in the
development of a vaccine that ideally could be administered in a single dose
and determined to be safe and effective. Many other factors complicate the
development of a worthy vaccine. The problems lie in the effect it has on
different age groups as well as the unborn, co-morbidity issues, and potential
latent effects. Once a vaccine is accepted to combat Zika, there lies the complexity of mass reproduction of the vaccine, the availability in poorer nations,
the safe handling of doses, stockpiling over long periods of time, and admin-
istrating the vaccine. Who decides the prioritizing of the vaccine, the pricing
of each dose especially when initial vaccines are likely to be limited. We face
some of the same problems with the pandemic of 2020. The attention given
to the Covid-19 virus has placed incredible pressures upon pharmaceutical
companies that are involved in several aspects of vaccine research, production, and testing for safety and effectiveness. December 2020 saw the initial
delivery and administered vaccination of the public began. Consequently,
until this pandemic can be controlled much of the work involved in therapeutic medicines and vaccines involving arboviruses will have to wait. The
parallelism of most viral diseases is obvious even to those who are just beginning to study, interpret data, and think about how viruses have both direct
and indirect consequences on people. A final thought is a reflection upon the
latest data that shows a lowering of Zika infected cases globally. The WHO
in their 2019 report states that the virus transmission is at its lowest levels
since the 2015-2016 epidemic. Of the 87 countries reporting the infection,
61 have populations of the *Aedes aegypti* mosquito.

The need for a Zika vaccine seems to have taken a back seat to more
pressing problems. New problems will continue to emerge in our world

beyond Covid-19, beyond malaria, and beyond known mosquito trans-mitted viruses until the next emergence of a novel virus epidemic begins to rear its ugly side and causes the illness and death of people. Eleventh hour medicine does not serve people very well. New, undetected, or ignored neglected disease outbreaks may go unabated until the toll taken is wide-spread enough to generate action. This is not meant to be a knock against our medical, scientific, and novel research studies, it is just that it is the way some things are treated. There is always room for improvement.

CHAPTER 16: ST LOUIS ENCEPHALITIS

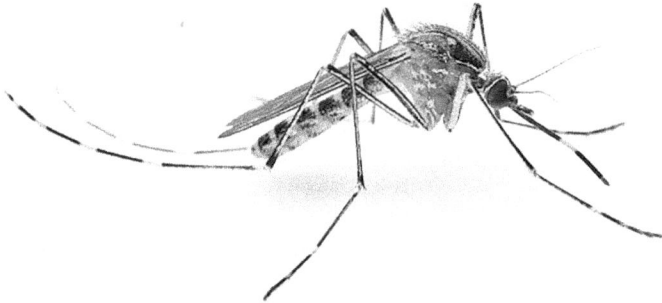

© MATT BERTONE 2015

Photo: *Culex tarsalis* primarily associated with the western states is a prominent vector of the St Louis virus. Note the middle of the proboscis that displays a white band, one of several identifying characteristics of this adult diapausing winter mosquito. Photo by Matt Bertone.

THE St Louis virus, a relative of West Nile virus originated from an epidemic of unknown sources in the fall of 1933 in St. Louis, Missouri. Recurrent outbreaks have occurred throughout the century in the Mississippi Valley and Gulf Coast. It is possible that migratory birds returning from endemic tropical habitats may be responsible, at least in part, for the enzootic transmission of the disease. It is also likely that there exists a yet to be known additional vertebrate or arthropod host serving as a local, over-wintering reservoir for the virus.

Today we know that the virus is transmitted from an enzootic mosquito-bird cycle to a mosquito-human cycle. The disease is often found in southern states year-round and in temperate climates anytime from late summer to early fall. Research suggests that an old, ancestral derived strain from South America has gone through a single point mutation in a single gene that translates an envelope protein giving rise to the human pathogen. This pathogenic development occurred about 116 years ago separating the North and South American strains.

There are currently 23 strains of the St Louis virus serogroup within the family *Flaviviridae* all consisting of ssRNA genomes. Introduction into the United States started with peri-domestic birds entering a new habitat as carriers of the novel virus into the region. Bird species that have tested positive and serve as reservoirs for the virus include finches, robins, blue jays, doves, pigeons, and sparrows. Wild birds appear to be the primary host for the virus but do not seem to develop the infection although a high viremic load is passed on to blood-feeding mosquitoes. The most prominent mosquito to transmit the illness is *Culex tarsalis*. In the East it would be *Cx. pipiens* or *Cx. quinquefascitis*, and *Cx. nigripalpus* in Florida. western states are infected by *Cx. tarsalis* and members of the *Cx. pipiens* complex. Resident mosquitoes of those regions pick up the virus from birds that had entered the new habitat and transmitted the virus to humans. Once again, humans serve as dead-end hosts where the disease can be found anywhere from Canada to Argentina. However most illnesses occur within the United States where epidemic outbreaks have been documented since the 1930's.

It takes anywhere from 5 to 15 days to develop symptoms of the illness. To determine the specific virus, antibody testing of blood or spinal fluid is used for diagnosis. Most people show no signs of the illness, however, those displaying symptoms will have a fever, headache, nausea, vomiting, and fatigue, all non-specific flu-like symptoms. Sometimes drowsiness is prevalent, and this symptom has led to the descriptive term - sleeping sickness. There are no specific treatments, therapeutic drugs, or vaccines on the market today. St Louis disease is considered rare, but it can cause encephalitic illnesses, and when severe, individuals may suffer from convulsions, paralysis, or even slip into a coma. Individuals increase their risk of getting the disease if they work or partake in recreational activities in known endemic regions. Older adults are at a much higher risk of severe symptoms from the viral infection.

From 1964 - 2005 there have been positive identifications of St Louis virus in 4651 individuals. From 2010 to 2019 there were 97 cases recorded in the United States with 6 deaths. In 2019 alone, 17 total cases were documented where 6 of those cases occurred in California and 9 occurred in Arizona. Mortality rates are anywhere from 5% to 30%, but higher rates occur in outbreaks where the population may be elderly and therefore the rates skew toward the higher percentages.

CHAPTER 17: YELLOW FEVER

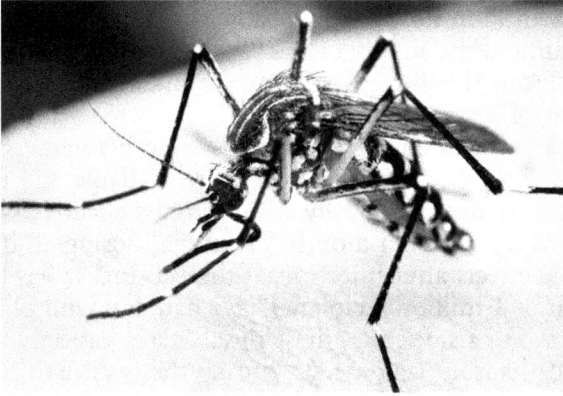

Photo: *Aedes aegypti*, the yellow fever mosquito and the transmitter of several other viruses throughout the Americas. Photo taken from the Marin/Sonoma Mosquito and Vector Control District, California web page.

YELLOW fever is one of the world's oldest diseases and has caused numerous epidemics throughout the centuries. The virus originated in Africa and was transmitted to the west through inter-continental shipping and the slave trade that started about 400 years ago. The first recordings of a yellow fever outbreak occurred in Barbados, Yucatan, and Cuba during the mid-seventeenth century. The illness spread uncontrollably on into Boston where the decision was made to quarantine ships arriving from the West Indies. Deaths mounted in Philadelphia and Charleston during the 1699 epidemic only to re-occur in 1732 in Charleston and New York.

In 1741, the term "yellow fever" was first used as it describes the symptomatic jaundiced skin coloration that affected the more severely ill. It was not until the later part of the nineteenth century and on into the early twentieth century that medical progress was achieved. In 1881, Carlos Finley identified the relationship of mosquitoes with yellow fever only to be ridiculed for his work. It was not until deaths occurred during the Spanish-American War and the destructiveness of the disease during the building of the Panama Canal that Walter Reed and his associates finally established the etiology of the disease from their research conducted in Havana, Cuba.

In 1936, the first live but attenuated (weakened) vaccine was first developed. The vaccine can be given to children as young as 9 months providing protection to that vulnerable age- group. Today, a single dose of the vaccine provides a lifetime of immunity. This was wonderful news except there are so many countries that have not received nor dispensed the vaccine to their inhabitants. For example, in 2016 Angola suffered from an outbreak where 2536 cases tested positive for yellow fever resulting in 301 deaths. The population had not been previously vaccinated and required over 11 million doses to help slow down and quell the epidemic. This severely limited the supply of the vaccine needed in many other areas of the world. From this limited supply, the practice of fractional dosing, about one-fifth of a standard dose was administered with similar results. The vaccine may remain effective for up to a year before needing a booster shot thus buying time to replenish the doses available and to provide for those in immediate need. Typically, a full dose is effective within 10 days with a 99% efficacy within a month. Fractional dosing all but eliminated any serious side effects after injection of the vaccine. It has been reported that only 1 out of 1 million recipients have had any kind of ill effect from the vaccination. On a side note, the United States has only had a handful of cases and deaths from yellow fever during the twentieth century. A case in 1996 was the first mortality since 1924 where the victim returned from Brazil having been infected by the virus while visiting and had not received a vaccination prior to departing the states. On a personal note, I was put on hold for several weeks in my need for the vaccine prior to leaving for Brazil in the summer of 2014. Interestingly only travelers to certain identified countries are advised to get vaccinated. Those countries, Africa, South and Central America, along with other vaccinations that are required or strongly recommended are listed for you on the CDC website for travelers.

The potential danger from the urban spread of the disease across borders has prompted the World Health Organization in 2017 to design, develop, and implement the Elimination of Yellow Fever Epidemic or EYE program. The program involves an agreement between 50 countries and has identified 40 at risk nations in Africa and the Americas. The goal is to eliminate yellow fever by 2026. It is best to note that only two viral diseases have been declared eliminated by WHO. The most notable is smallpox, although the United States and Russia maintain a secure, live culture of the virus, no known illness has been observed. The second virus eliminated from the planet was rinderpest, a cattle disease. These virus diseases have been eliminated through the development, distribution, and implimented vaccination programs. Even polio still lingers in Afghanistan and Pakistan while the virus is nearly gone throughout the rest of the world. The younger audience reading this book never had to experience the long lines where polio shots (plus one booster) were given over a period of several months. Hey, as a child, you got a free lollipop for your cooperation. I cannot help but wonder if a lollipop might influence the anti-vaxxers today.

The yellow fever virus is a member of the *Flavivaridae* family of arboviruses. It is a relatively rare illness in the United States today primarily because of mosquito abatement programs targeting the mosquitoes most responsible for disease. In addition the developed vaccine provides a life-time immunity. Yellow fever, like so many of the arbovirus illnesses, displays non-specific symptoms or asymptomatic conditions in most people. The developing symptoms usually occur within three to six days following transmission allowing time for the virus to replicate within the human host. Serious infections of the hemorrhagic nature of the disease occurs about 24 hours following a brief pause of what appeared to be a mild symptomatic infection. Symptoms become extreme where organs begin to fail, especially the liver, hence the resulting yellow skin or jaundice. Bleeding occurs where the severely infected person vomits a blackish fluid from the stomach. These severe symptoms often lead to anywhere from 30% to 60% mortality. It may even be higher in poorer nations that cannot provide medical service during an outbreak. WHO has designated 34 countries in Africa and 13 countries in the Americas as being endemic for yellow fever. These decisions were based upon modeling data, at times from under-reported cases in Africa and the Americas. In 2013 for example, it was estimated that there were somewhere around 84,000 to 170,000 known cases of yellow fever. Mortality was estimated to be between 29,000 to 60,000. Africa has 90% of all yellow fever infections in the world where 200,000 cases occur per year with 30,000 deaths not including extreme outbreaks of the disease. The greater the population density of people coupled with an increase in concurrent mosquito density places the human-mosquito contact at an extremely high percentage. Exposed viremic humans for up to five days following the initial transmission, and the potential for a new blood feeding episode, may likely pass on the ability of conditioning mosquito carriers that may ultimately lead to local epidemics of yellow fever. In the fall of 2020, an undiagnosed outbreak occurred in Guinea that corresponded with Covid-19. These infections corresponded with infections of Lassa fever, a combination that complicated the diagnostics of the illness. In Nigeria, over a 5- week period, 52 cases followed by 14 deaths occurred. Of those infected, 31 people were unvaccinated, 18 others were of an unknown status, and 1 case claimed vaccination but was unable to produce a government issued vaccination card. Later, 2912 children received the yellow fever vaccine following confirmed immunoglobulin positivity. Only 40% of the population has received the vaccine as of 2019, a level far below the needed herd immunity that a vaccinated population needs to achieve.

The yellow fever mosquito, *Aedes aegypti* is most responsible for epidemics over the recorded history of the disease throughout the Americas. Depending upon where the epidemic may emerge in the world, several mosquito's species from the genera *Haemagogus* and *Wyeomyia* are also known carriers of the virus. Transmission of the virus has been identified from three cycles. The first is the sylvatic or jungle cycle involving the relationship between the mosquito and non-human primates (monkeys). In Africa, one of the mosquitoes commonly involved in the sylvatic cycle is *Aedes africanus*. A second cycle involves a mosquito-human relationship specifically with those people who live or work in areas that border the jungle environment and have contact with mosquitoes carrying the virus. The third cycle is the urban cycle where humans and mosquitoes (*Aedes aegypti* and *Aedes albopictus*) pass the virus in a classic arbovirus manner. Viremic humans are fed upon by mosquitoes that are associated with sylvatic or savannah habitats. It is in these habitats where the mosquitoes have adapted to urban life provided for by ample numbers of artificial containers selected for oviposition. Along with these dwellings, an ample pool of humans for blood feeding satisfies the nutrient requirement for egg maturation. Repeated cycling, a pattern that must be interrupted to control the mosquito to human to mosquito arboviral repetitiveness. Vaccines, therapeutics, mosquito control strategies, and self-protective measures are keys to success in fighting any mosquito vectored disease.

CHAPTER 18: CHIKUNGUNYA

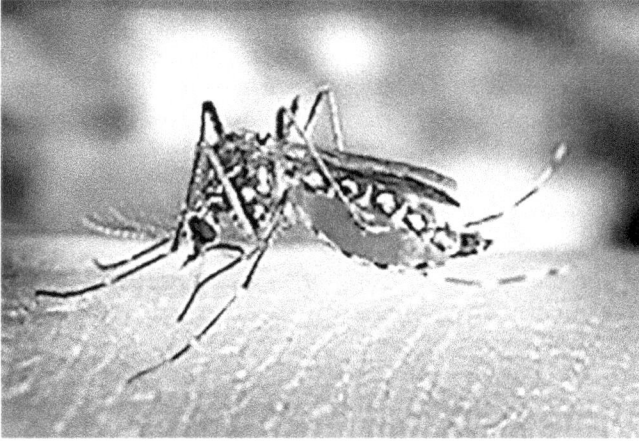

Photo: *Aedes africanus* photocopied from the Wikipedia website. This mosquito can transmit yellow fever, Zika, dengue, Rift Valley, and chikungunya disease in tropical Africa. It is a sylvatic or jungle cycle mosquito that inhabits natural treeholes.

CHIKUNGUNYA was first isolated in 1952 in Tanzania. It remained as an isolated illness with small outbreaks in regions of Africa and Asia. The illness was treated as a neglected tropical disease until rapidly spreading outbreaks began to occur in 2004 involving 60 countries of Africa, Asia, Europe, and the Americas. Two years following the rapid spread, WHO reported that one third or 500,000 cases of infection were documented on the Reunion Island. It later spread to India where 1.5 million were infected and continued to spread from infected travelers throughout Asia. By 2007, Europe and especially Italy reported infections possibly linked to *Aedes albopictus* while travelers from Europe to the United States and Taiwan spread the disease. Larger outbreaks began to surface in the Americas and Europe where in 2014 over 1500 cases were identified. The spread was so rapid that in 2015, along with the Zika outbreak, 693,489 suspected cases were reported in the Americas.

In the previous chapter a parallel was made between the symptoms of Zika and dengue. Chikungunya can now be added to the list sharing similar symptoms even though the virus is from a different family and does not display encephalitic characteristic of EEE and WEE associated with the *Alphavirus*. Fever, rash, muscle and joint pain, headache, nausea, and fatigue share the symptoms of viral infections brought on by the viruses transmitted by mosquitoes. There are reports of isolated, yet significant severe symptoms occur when infected by this virus. The cases where asymptomatic infected individuals, roughly 80%, was common in Zika and dengue infections, chikungunya may well be symptomatic in 75-80% of the infected. Joint pain can be debilitating; hence the derived interpretive name is associated with chikungunya disease, that is, 'to be bent over' or 'to become contorted' derived from the Kimakonde language. Once again, co-morbidity in a person can exacerbate the symptoms and the length of time required to heal. Like so many of these viruses, no vaccine or therapeutic drugs have been developed. Treatment of symptoms, no different than treating cold and flu symptoms, is the only treatment to provide some relief. WHO reports that there are several vaccines being considered but are only within the first stages of trials testing the safety and efficacy of the vaccine. Consequently, the licensing of a vaccine is several years away. This is not unusual in the development of vaccines and it is astounding how quickly several vaccines are being reviewed for emergency treatment and public deployment in treating the Covid-19 virus. The key for Covid-19 vaccine development is the genetic modification of a protein or proteins in the spikes of the Covid-19 virus that is quite different than the attenuated virus approach commonly associated with vaccine development. The shift from dead or attenuated viruses in a vaccine to the selection of mRNA translation coding designed to trigger antibodies apparently has revolutionized the vaccine developing timeline from years to months.

The under-reported number of chikungunya cases, as well as under-reporting of Zika and dengue, makes it nearly impossible to determine the rate of infection in so many parts of the world. Here in the United States where medical problems are reported on a regular basis to state health agencies numbers have shown a dramatic decrease in this disease. For example, in 2014, 2811 cases were identified within the continental U.S. Twelve were from local transmissions, all in Florida while all the others were from travelers entering the U.S. By 2020 there have been no local transmissions or travelers testing positive for the virus. This holds true for all U.S. territories of Puerto Rico, Virgin Islands, and the American Samoa that recorded 4710 cases of which 4659 were locally transmitted.

The virus is from the family *Togaviridae* and the genus *Alphavirus*. It is a ssRNA virus where both vertebrates and non-vertebrates serve as hosts or act as a reservoir for the virus. Implicated as reservoirs include non-human primates (vervet monkey and baboon), rodents, birds, and small mammals that play a role in possible re-emergence of outbreaks following a period of inactivity in humans. In Africa, the mosquito transmitters involved include *Ae. africanus, Ae. luteoscephalus, Ae. opek, Ae. furcifer, Ae. taylori,* and *Ae. cordellieri* found throughout savanna and forested habitats. The more recent outbreaks in urban areas include *Ae. aegypti* and *Ae. albopictus* species that pick up a sufficient viremia from infected humans and then transmit the virus to other humans. In some regions of the world, especially in the cooler temperate biomes, *Ae. albopictus* is the predominant spreader of the disease.

The virus enters the cytoplasm of the cell, reproduce, and the newly developed virions escape the cell membrane and enter adjacent cells to further reproduce. The onset of the illness in humans is anywhere from 2 to12 days. Often displayed initially as a fever, some symptoms may become chronic. This illness is rarely life threatening while most individuals can recover completely. Like other arboviruses, the person is likely to be immune to any further re-infections.

Transmission begins when a naïve mosquito, one that does not harbor the virus, takes a blood meal from an infected individual that is viremic. The virus enters the mid-gut of the mosquito, can replicate itself, and eventually finds a path to the mosquitoes' salivary glands. At this point the virus, through mosquito blood feeding, passes the virus into a naïve human. Interestingly, the reproduction of virions in this second transfer amplifies the virus into a relatively high concentration within the human cells. From there, the mosquito picks up the amplified viral load and begins the cycle all over again. The infected mosquito can transmit the virus for life.

It is well known that *Aedes aegypti* and *Aedes albopictus* are the two most prominent mosquitoes involved in the transmission of chikungunya. *Ae. albopictus*, being cold tolerant as compared to *Ae. aegypti*, has afforded the species the ability to extend its range beyond the tropical and sub-tropical regions of the world. It has been able to establish populations in urban, and suburban temperate climates and is known to establish populations in shaded areas associated with parks where human-mosquito contact ratios increase. In addition to these two species, there are those species associated with the *Ae. furcifer-taylori* group and *Ae. luteoscephalus* species. These species are found in sub-Sahara Africa and are zoonotic vectors of several viruses which includes chikungunya.

CHAPTER 19: EASTERN EQUINE ENCEPHALITIS

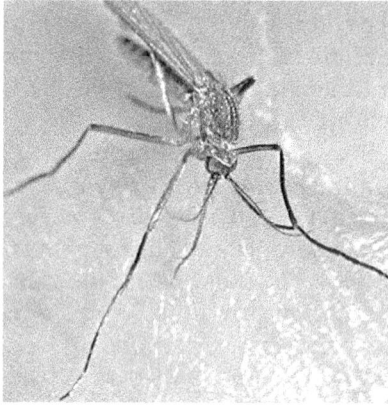

Photo: The black tailed mosquito, *Culiseta melanura*, live in and around swamp-like habitats. Photo credit goes to C. Roxanne Connelly, University of Florida.

ENCEPHALITIS is an acute inflammatory disease that involves the brain and spinal cord. Most cases are caused by a virus and sometimes caused by bacteria, fungi, and autoimmune disorders displayed in infected people. The incidence or infectious rate in the United States for EEE is one case for every 200,000 people. To compare, herpes simplex virus is the most common viral illness where 12-13% of individuals in their late teens or middle aged have HSV 1 or HSV 2. During outbreaks, the number swells to about 50% in epidemic proportions. Arboviruses are the cause of 10% of all the infected cases while during a regional outbreak the number of infected may rise as high as 50%. There are five encephalitic viruses that are monitored in the United States. Eastern equine encephalitis (EEE), Western equine encephalitis (WEE), St. Louis encephalitis (SLE), LaCrosse encephalitis (LAC), and West Nile virus (WNV).

This chapter is devoted to a serious illness where the death rate can be as high as 30-35% and may exceed 50% during an outbreak of Eastern equine encephalitis. The illness is a part of the virus family *Togaviridae*, and the genus *Alphavirus*.

In 1831 in the state of Massachusetts, 75 horses were determined to have died from a mysterious illness. The first known human fatality occurred in 1838 in New England. Over time it was determined that the mosquito-bird or mosquito-reptile cycle made the leap to the mosquito-to-horse-human cycle. A large-scale outbreak occurred in the Mid-Atlantic where 25 of the reported 34 cases of human infection proved to be fatal. Survivors face the possibility of ongoing neurological difficulties. The disease is predominantly found in the eastern and Gulf States as well as within regions of the Great Lakes.

The mosquito associated with EEE is *Culiseta melanura* that displays a preference for avian blood. The habitat is a swampy, freshwater wetland containing hardwoods. Transmission of the virus occurs primarily in the early spring on through to the fall. The disease is prevalent in the Gulf States even through the winter months. Other mosquitoes that could transmit EEE to dead-end horse and human hosts are *Coquillettidia perturbans*, and a variety of *Aedes* and *Culex* species. There are other mosquitoes that may serve as bridge vectors. *Ae. sollicitans* inhabitants of coastal areas and *Coquillettidia perturbans* that inhabit inland habitats. These mosquitoes feed on birds in the spring picking up the virus and then passing it on to other mammals. Research has also suggested that the virus occurs in overwintering species, specifically pointing to reptiles. Migratory birds returning from their winter destinations from southern endemic areas contribute to the spread by carrying the virus back to the colder regions.

Although considered a rare disease in humans, infrequent contact of infected mosquitoes and humans occurs because the swamp environment limits access and therefore contact. In 2020, the CDC reported that there had been only 9 fatalities. The average number of cases is 11, however, 2019 was exceptional with 38 fatalities mostly in Massachusetts and Michigan. From 2010 to 2019, 107 cases were reported to the CDC where 48 were fatal. That is a fatality rate of 44%, exceeding the often reported 33% incidence. The most vulnerable appear to be those younger than 15 and older than 50 years of age. Survival of the infection seems to confer a life-long immunity. However, like so many other arboviruses, there appears to be no cross-immunity protection against other *Alphaviruses* (e.g., WEE), *Flaviviruses* (e.g., WNV), or *Bunyaviruses* (e.g., LaCrosse virus).

Once infected from a blood-feeding mosquito, it takes anywhere from 4 to 10 days before the onset of symptoms. Asymptomatic individuals exist but the meningitis effect depends upon the age of the infected individual and a variety of other factors such as co-morbidities. The initial onset of symptoms includes a fever, chills, malaise, arthralgia, and myalgia that may last up to 2 weeks. The more serious effect is the neurological disease that effects older children and adults following several days (2 to10 days) from the initial onset of symptoms. Symptoms include headache, vomiting, diarrhea, seizures, behavior change, drowsiness, and coma. Those who recover from EEE may require long term care and continue to display mild forms of brain dysfunction, personality disorders, seizures, paralysis, and central nerve damage. In severe cases, the older adults often die within a few years.

Currently there is a vaccine for horses that was developed several years ago. There are no approved vaccines or drugs to treat humans. A human vaccine was developed by the U.S. Army Medical Research Institute of Infectious Disease (USAMRID) at Ft. Detrick, Maryland in the 1980's. The vaccine was developed to provide protection for those deployed in known EEE zones. The use of the vaccine was halted by the FDA for not following the guidelines required for the deployment of a vaccine distributed to researchers because of unregulated trials. Researchers working on EEE may still receive the vaccine, but it comes at a great cost to individuals. A fee of up to $20,000 must be paid, the final cost dependent upon the study involved and the individual must go to the USARMID facility to receive their vaccine. Following the inoculation, the individual must return periodically to partake in a physical and have blood samples drawn. One other important factor is that the vaccine seems to be very short-lived, and a repetitive booster would be required for those involved in EEE research. Finally, there exists a cost-benefit downside to the production of the vaccine, costing millions of dollars, to serve less than a handful of infected individuals. For argument's sake, let us say 100 individuals require the vaccine each year while by comparison, individuals suffering from Lyme disease are 30,000 or more each year. To conclude, the expense to control a rare disease like EEE to such a few patients is difficult if not impossible to justify.

CHAPTER 20: WESTERN EQUINE ENCEPHALITIS

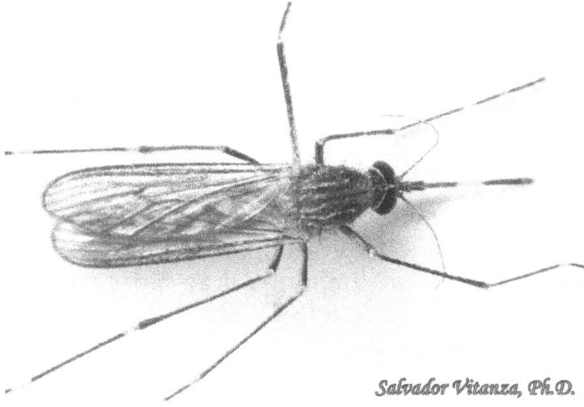

Salvador Vitanza, Ph.D.

Photo: *Culex tarsalis* a transmitter of Western equine encephalitis.
Photo by Salvador Vitanza, PhD, taken in El Paso County, Texas.

WESTERN equine virus is in the same viral family, *Togaviridae*, as is chikungunya and Eastern equine encephalitis discussed in previous chapters. Along with WEV, CHIKV, and EEV other new world diseases of the Western Hemisphere include Venezuelan encephalitis. The viruses in this family are ssRNA types with only two known genera. Rubivirus has only one species that causes rubella also known as the German measles or the three-day measles. Unlike most arboviruses, this illness is passed from person to person through airborne expectorants, mucosal contact, and from mother to the unborn through the blood stream. It is often a mild illness lasting about three days however some serious complications may result especially in the unborn. The MMR vaccine (measles, mumps, and rubella) is available to most individuals lessoning the spread and ill effects of the WNV.

The genus *Alphavirus* consists of 29 species of virus of which several are related to mosquito borne infections. Most cases of WEE occur in birds and rodents, a zoonotic disease where these vertebrates serve as reservoirs of the virus. It is the unvaccinated horses that serve as sentinels for the disease where the cycle of enzootic transfer continues, and the virus is always present. Most humans are asymptomatic and there is apparently no rash associated with the infection. However, symptomatic individuals share the same non-specific viral conditions as has been repetitively expressed in previous chapters. Most of the serious neurologic disorders occur in the noticeably young and older population. Children may suffer from permanent disabilities (seizures), in as many as 50% of those infected, while the older population of infected adults face death from complications associated with the disease with a 5% to15% fatality rate. Those recovering from the initial illness may experience years of acute neurological disorders such as fatigue, headaches, and irritability. The calculated rate of incidence is 1:1000 for adults, 1:58 for children one to four years of age, and a 1:1 ratio in infants. All things considered, WEE is a less severe illness than EEE with an overall human fatality rate of 3%. It remains much higher in horses. The good news is that the incidence of disease has been declining since the 1970's likely due, in part, to mosquito control abatement strategies and the decrease in mosquito-human contact.

It may well be that the western form of this disease was predicated upon the mutational advancement offered through an ancient recombinant alphavirus, notably the EEV genome establishing the WEV. The virus was first isolated from the brain of a dead horse from an outbreak in the San Joaquin Valley, California in 1930. The establishment of WEEV in humans as the causative pathogen was determined after the death of a youngster in 1938. A large outbreak in the United States occurred in 1941 where over 3000 people were confirmed to have been infected. Since 1964 only 700 cases have been confirmed. Most of these cases exist west of the Rockies, west of the Mississippi, and in selected regions of California and western Canada. Portions of South America and especially Argentina have recorded outbreaks of WEE. Wherever there are warm temperatures and heavy rains, infections increase in the summer months that follow. This takes place when the mosquitoes that transmit the virus have increased their populations because of preferred habitat conditions, hence their ability to increase their transmission capabilities. Transmission may occur from April through September with peak times being July and August. Males are the primary recipients of infection usually twice that of females. This is likely because of outdoor activities and occupations that males are more often associated with.

The mosquito most often involved in the transmission of WEE is *Culex tarsalis* that serves as the enzootic, epizootic, and endemic transmitter of WEEV. It seems that the existence of the virus as an endemic disease is associated with two cycles involving six viruses forming the WEEV complex. The first cycle involves several species of birds along with nestling bird species that serve as amplifying hosts. The second cycle involves jackrabbits as hosts and *Aedes melanimon* or *Aedes dorsalis* as transmitters in western states like California and even as far east as Colorado. When a human outbreak occurs, infections are also associated with mules, horses, pheasants, several rodents, reptiles, and several bird species. The mosquito starts out feeding primarily on birds but as the summer approaches they make a switch to mammals. Transmission of the virus is transferred from infected mosquito to human and horses that serve as dead-end hosts. It is not from aerosols or human expectorants that spreads the virus, but it has been shown to occur between the infected mother and fetus. Blood transfusions seem unlikely because of a low viral load but may be possible.

Once infected by the blood-feeding mosquito the virus is passed into the subcutaneous tissue, and replication of the virus begins to produce RNA and associated proteins that eventually find their way to the lymph nodes. The potential viremia, if the viral load is sufficiently high, enters the central nervous system by crossing the blood-brain barrier resulting in neurological inflammation and potential lifelong debilitation or death.

CHAPTER 21 : LA CROSSE ENCEPHALITIS

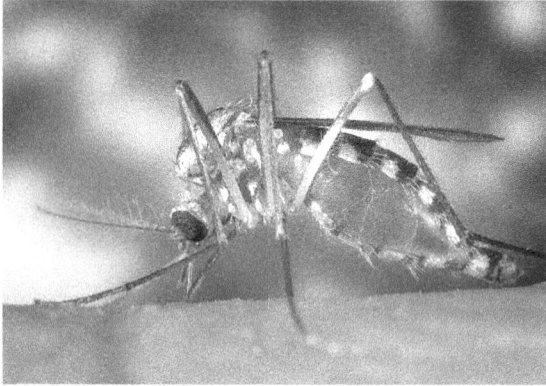

Photo: The Eastern Tree Hole Mosquito, *Aedes triseriatus*. It is difficult to separate identification between this species and the closely related *Aedes hendersoni*. Transmitters of La Crosse virus. Photo taken by James Gathany/CDC.

IT was 1960 when the first fatality from La Crosse virus was identified. A 4-year-old girl from La Crosse, Wisconsin led researchers to the discovery and ultimate research into another novel appearing virus that has been described as an endemic disease in the United States. Thousands of individuals have been infected since 1960 where the more severe cases fall into an average yearly range of 70-130 cases a year.

The mosquito *Aedes triseriatus*, the eastern tree-hole mosquito is the main transmitter for this virus. This mosquito is a forest dwelling species whose larvae may be found in both natural and man-made containers. The mosquito prefers small mammals to feed upon including the eastern chipmunk and the eastern grey squirrel. These animals serve as amplifying hosts while the mosquito displays transovarial or vertical transmission of the virus, only later to spread the virus through its blood feeding behavior to its dead-end host, humans. Dead-end hosts are so designated when the host, in this case individual humans, are incapable of producing a high concentration of the virus to infect mosquitoes. The reliance of transmission is relegated to the reservoir hosts to continue the cycle. Recently, studies involving the invasive mosquito *Aedes albopictus*, suggest a shift in the viral ecology that may lend itself to a new strain of the virus. In both species, the ability to over winter as eggs allows for the vertically transmitted virus from adult female to her unhatched eggs to survive the cold and emerge in the spring capable of transmitting the virus once maturity is reached.

Most infected individuals have no symptoms although severe cases have been documented. Unlike St Louis virus where the more severe cases affect primarily the elderly, the LaCrosse virus has an opposite effect. Serious cases occur in children under the age of 16 where they may experience seizures, convulsions and even paralysis.

The eastern regions of the United States most likely to report cases of the illness are from the upper midwest, Mid-Atlantic, and southeastern states. The warmer Golf states may report cases during the winter months although most infections are reported to be during the late summer and early fall times of the year. There is an average of 63 to 68 cases reported on a yearly average. However, this seems to be an under-reporting problem as well as an under-diagnostic problem primarily because the of the asymptomatic or mild symptomatic results of infection.

Individuals will develop symptoms within 5 to 15 days after contact with an infected female mosquito. Symptomatically, this virus, without diagnostic testing to affirm the viral type, is part of the collection of non-specific viral disease symptoms. Furthermore, because so few are severely impacted by the virus, a negative cost/benefit investment hinders the development of drug treatments specific to the disease as well as the development of a vaccine.

The number of cases reported to the CDC during the period from 1964 to 1995 accounted for over 2200 known illnesses with a certainty that this reflects underreported cases . The projected mortality rate based upon the past decade of records is less than 1%. From 2010 to 2019 there have been 741 cases with 8 fatalities. The range of positive cases over this time was a low of 31 cases and a high of 116 case. In 2019, the data shows a total of 55 cases where 26 of those occurred in Ohio, 5 in North Carolina, and 12 in Tennessee.

The La Crosse virus is part of the California serogroup made up of over 50 virus species including the little-known Jamestown Canyon virus and the Snowshoe Hare virus specifically associated with the United States and Canada. These viruses belong to the family of viruses known as *Bunyaviridae*, genus *Orthobunyavirus*. The La Crosse encephalitic virus is probably the most important of the California serogroup of viruses. Interestingly, the mosquito itself may serve as a reservoir host without first feeding upon a viremic host. Furthermore, it seems that the male mosquito, through the transovarial mechanism, passes the virus from male to female in the wild through mating, a venereal transmission. Finally, my home state of New York has shown that the mosquito *Aedes canadensis* may also serve as an enzootic contributor to the sylvatic cycle for the transmission of La Crosse, but this has not been fully developed and therefore not well documented for needed clarity.

Chapter 22: Jamestown Canyon Virus

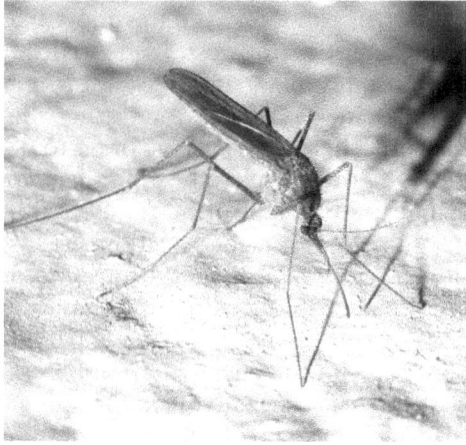

Photo: *Culiseta inornata* known as the winter mosquito or winter marsh mosquito. Photo shows the mosquito on a bed of snow. Photo compliments of Wikipedia, the free encyclopedia.

I thought I would introduce to the reader two unheard-of viral illnesses that are somewhat unique in that they are limited geographically to the United States and Canada. Most of the arbovirus diseases discussed in the previous chapters are well-known and often capture the attention of the news media. Disease breakouts that lead to serious epidemics associated with localized hotspots are presented to the public through the cable news network's where even the most casual of interested people become aware of these viral diseases. To my knowledge, Jamestown Canyon virus and Snowshoe Hare virus has never made the big story on the evening news.

Jamestown Canyon virus and Snowshoe Hare virus belong to the same arbovirus family as the La Crosse virus and are members of the California encephalitis serotypes (*Bunyaviridae*: *Orthobunyavirus)*. As expected, these diseases are often asymptomatic or so mild that the actual number of infected goes undiagnosed and therefore under reported to the CDC. These two diseases are encephalitic with exceedingly rare severe cases leading to few mortalities. There are no vaccines or therapeutic drugs to combat the virus. Treatments are used to alleviate the symptoms of fever, headache, coughing, runny nose, sneezing, sore throat, and fatigue that will develop within two days and up to two weeks after the infected mosquito contact had occurred. It is rare for anyone infected to show signs of encephalitis or meningitis. Death is also rare for these diseases. Approximately half of all infected people are hospitalized.

Jamestown Canyon virus was first isolated in 1961 in the state of Colorado. The mosquito that was first to be recognized as a carrier and transmitter was *Culiseta inornata* although several other mosquito genera have since been confirmed in *Aedes*, *Culiseta*, and *Anopheles* species. It is unclear that even though some of these other species of mosquitoes are carriers, the question remains as to whether they can transmit the virus to humans. The virus has been isolated as far north as Alaska and ranging south and east across Canada into New York and New England with the southernmost range into the Mid-Atlantic coastal states. The primary mosquito genus is that of the *Aedes* whose reproductive cycle is univoltine relying upon springtime vernal pools and rock pools for their reproductive success. The spread of the virus is in the early spring and into the fall prior to the onset of cold temperatures. The mosquitoes are capable of transovarial transmission using deer as their vertebrate host. It has been reported that humans do not produce enough of the virus to reinfect the next feeding mosquito that would then be responsible for the spread of the disease much sooner. The ecology of the species, the time of year, the geographic location of the mosquito species, and contact with its vertebrate host collectively must mesh for any significant number of people to be infected. The mean number of cases is around 15 each year. Underreporting of the less severe cases limits the actual number. The states of Minnesota and Wisconsin have reported about half of the last three years (2017-2019) infections. During the time frame of 2010 - 2019, the CDC has reported 225 cases and 5 deaths. In 2019, Minnesota reported 54 cases while Wisconsin reported 84. The recent increase in reported cases may be related to several changes in policy and climatic conditions. The CDC suggests that an increase in awareness of the illness along with the 2013 policy of antibody testing for the virus has led to the reported increase in cases. Furthermore, an increase in select species of mosquitos like *Cs. Inornata* along with changes in climate have extended the longevity of the mosquito increasing contact with infected deer and later transmitted to humans.

CHAPTER 23: SNOWSHOE HARE VIRUS

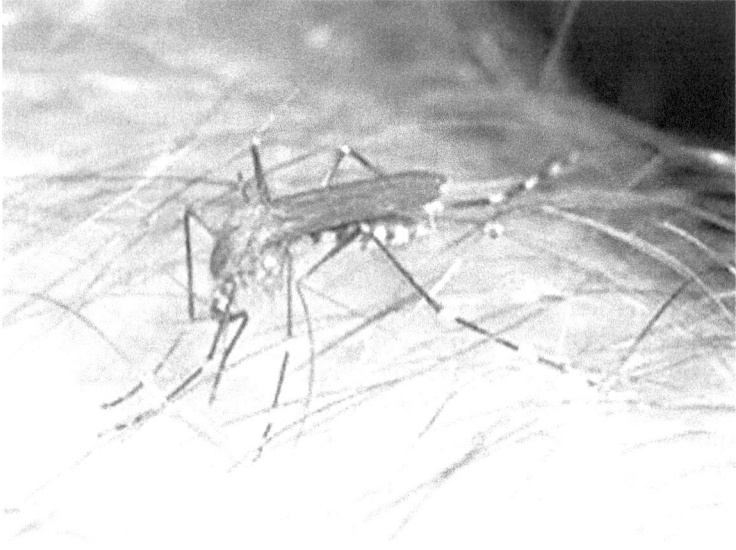

Aedes canadensis, one of several mosquitoes to transmit SSH virus.

THE Snowshoe Hare virus is an even more rare event where humans are concerned. The first isolation of the virus occurred in Montana in 1958. The illness in humans was first recognized in 1978. The SSH virus has been found throughout Canada, Alaska, and several states that border Canada.

It was established that the Lagomorphs (hares and rabbits) are the enzootic vertebrate hosts that supply the mosquito with the virus. There are several other mammals and birds in addition to the hare that have served as amplifying hosts.

The virus is classified in the family *Bunyaviridae* and in the genus *Bunyavirus*. The virus is a single stranded RNA virus of the California serogroup that includes Jamestown Canyon virus. Several species of mosquitoes in the genera of *Aedes*, *Culex*, and *Culiseta* have been associated with the disease. It has been noted that the 'spring mosquitoes', those that emerge in vernal pools and forested habitats are associated as well with this virus. The virus can survive the long and cold winter months and the infected female passes the virus along to its progeny.

This is an extremely rare illness that is likely underreported. Those who develop symptoms match the same non-specific characteristics associated with colds and flus. The ill may develop some encephalitic type symptoms. No vaccine or drugs have been developed to treat this virus.

CHAPTER 24: MOSQUITO CONTROL STRATEGIES

WHEN there are limitations or non-existence of therapeutics and vaccines to control diseases transmitted by the mosquito we are left with the only recourse that has been adopted and put into practice for several decades, vector-control of mosquitoes using insecticides. Control occurs when the overall number of an isolated population of known mosquito transmitters is reduced or eliminated. Success in the use of insecticides is losing ground to insect resistance. Climate change allows for the expansion of the mosquito's range into new territory, and the introduction of novel pathogens are reasons to develop alternatives to what has been the practice of the often-uncontrolled use of insecticides.

The very first strategy employed by humans occurred millions of years ago and began through the practice of flailing one's arms to ward off the annoying buzzing sound in their ears. This response quickly progressed to the 'swatting technique' still employed today. Let us all remember that the correlation between mosquito blood feeding and the transmitting of disease was not fully recognized until the late nineteenth century even though the need to reduce or eliminate these pests reigned supreme at a much earlier period in human history.

The greatest progress in mosquito control has taken place from the early and mid-twentieth century to the present. Most researchers recognize that there exists a choice between vector control and biological control of mosquitoes. Vector control has been around for decades and includes strategies that involves the use of air or ground applications of insecticides, the draining of stagnant wetlands, and adding larvicide dunks to vernal pools. These practices served as vector controls often applied on a large scale. Such large-scale approaches to vector control often have severe limitations as to the effectiveness of those strategies. On a much smaller scale, protections are employed for personal use and home improvements. Alternative strategies included adding screens to windows and doors, the introduction of air conditioners and, believe it, even television. The advice to drain flowerpots or other water holding containers around the home also helps in controlling mosquito abundance. Personal protection requires body sprays containing DEET, clothing sprayed with permethrin, and wearing long sleeve and long pants if you can withstand the summer heat. All these methods provide some protection from the blood seeking female, but it apparently is not sufficient to ward off outbreaks of local epidemics, the emergence of novel pathogens, and the devastating illness and death that accompanies these outbreaks. In all fairness, vector control does have some success. During World War II the use of dichloro-diphenyl-trichloroethane (DDT) was applied as a body dusting (Italy, 1944) to eliminate body lice and the typhus illness associated with the insect. The use of DDT had a positive effect upon eliminating malaria bearing mosquitoes from most of the industrialized world and had a similar effect on so many other mosquito-borne diseases. In 1962, Rachel Carson published her book *Silent Spring*. In the book she pointed out many of the downsides in the use of insecticides for agricultural control of insects and linked the practice with numerous environmental problems. It took 10 years for the United States to ban the use of DDT and several other environmentally unfriendly chemicals. Carson has since been blamed, by notables like former senator Tom Coburn and novelist Michael Crichton, for the deaths attributed to malaria and other illnesses and suggest that she had these chemicals banned from use. These accusations are far from the truth, but a sad chapter in the history of science and society.

Today, the use of insecticide treated bed nets and insect application to the inside of homes in malaria infested areas has decreased by half the number of illnesses and deaths. However, there are always draw backs to any strategy. For instance, insecticide resistance is on the rise, repetitious application is a costly requirement, while new bed nets need to continually replace the older, least effective, worn out nettings. Unforeseen was the use of bed nets, taken from homes, to serve as fishing nets by the local fishermen thus polluting local bodies of water and having a deleterious effect upon the local economy and well-being. Finally, the cost-benefit of implementing any strategy is difficult to attain, especially in the poorer of nations or in nations associated with neglected tropical diseases.

Alternative strategies involving biological control methods or combined vector and biological methods are a part of the integrated vector management programs or IVM. Allow me to stress, through my repetitive comments, the need for alternative methods being driven by the overwhelming problem of insect resistance (IR), and the changes in climate resulting in extending the range of some mosquitoes. Concern for the introduction of novel pathogens all lead to the increase of contact between the mosquito and the host. The spread of a disease by several mosquito species poses an entirely different problem. West Nile virus initially had been associated with a few *Culex* species of the *Culex* complex, both as transmitters and as carriers. Today, that number has climbed to over 60 species of mosquitoes in North America. Will this co-existence of mosquito and virus develop into a greater number of potential transmitters and not just found to be non-vectoring carriers? What about other diseases in other countries? The need to expand the tools in the toolbox of mosquito control choices is forever growing as the mosquito, the pathogens, and the increase in human-mosquito rates of contact continue to create a global threat.

Public health departments and the agricultural community have accepted and continue to rely upon vector controls over the growing number of environmentally safer biological control methodologies. Applications for the use on large scale treatments often fail or are limited in use for several reasons. It may be as simple as failure to effectively implement the distribution of insecticide, sudden wind or rainstorms diluting or carrying the application to the wrong area resulting in ineffective target coverage. Add the growing result of insect resistance, there are practical economic concerns in terms of financial short falls, political interference, or even public outcry to limit or halt a program. The fear of pollution runoff, the polluting of local drinking water (the result of fracking as an example), or the fear of human contact with a little understood chemical effect upon humans and the environment all pose problems within communities.

Relatively new thinking is to scale the strategies as a problem of niche products that eliminate many of the ineffective barriers witnessed throughout so many of the vector control management projects. This requires knowledge of the target species of mosquito, their biology, behaviour, and ecology. These needs leave an open door for the aspiring mosquito researcher to investigate on a scale designed for local niche requirements and perhaps provide answers to the descriptors needed to manage these transmitters of disease whether they be vector, biological, or a combination of strategies.

One of the more interesting techniques involves a process known as gene drive. The creation of genetically modified mosquitoes and the Release of Insects carrying a Dominant Lethal gene (RIDL) strategy. Male mosquitoes are equipped with a lethal transgene that effectively creates sterile progeny. The males are lab produced in the presence of tetracycline as a suppressor and they survive only in its presence. Once released into the wild, the first mosquitoes, tagged as OX513A, mates and dies because the tetracycline is not available in the wild. These genetically engineered mosquitoes were put to the test in the Grand Caymans using transgenic *Ae. aegypti* with great success. The risk for any of the GM products involves consideration of the cost-benefit, safety, non-target effects, adverse effects on biodiversity, consequences of gene flow, and environmental changes. Allow me to point out that the application was under a smaller, controlled environment. That being an island and not an overly expansive territory as discussed earlier.

Clustered regularly interspaced short palindromic repeats, or CRISPR, is a form of genome editing for organisms that reproduce sexually. A typical allele has a 50% chance of being inherited. A gene drive modified sequence has a 90 % chance of inheritance and offers a vast array of applications. For example, this control strategy applies to plants where their genome is modified to the sensitivity of herbicides designed to eliminate invasive organisms. The addition of a genetically altered mosquito, modified to resist the malarial parasite, is a welcomed addition of a new and perhaps more effective tool to control pest species. This method of modification and control is less costly than other methods. It is a targeted, species-specific technique, and is less controversial than the use of pesticides in our everyday lives. Drawbacks to the use of this technology include the potential for GMOs to escape into the environment prior to the completed study, the potential for cross breeding, changes in gene flow in specific populations, and ultimately the potential of adverse effects upon the environment.

A precursor to gene drive is the use of the sterile insect technique or SIT. This method of control involves the release of sterile males into the environment having been sterilized through the application of irradiation or chemicals. In theory, sterilized males, in great number, are released into the target area, notably in a small and confined application site for better control. They mate with their female counterpart and result in the blocking of fertility, therefore reducing the population of that specific organism. Repeated sterile male release lowers the female population and therefore the overall abundance of the target species. In effect, over time, the treatment reduces the threat of transmission by reducing the mosquito-human contact ratio. Total loss of a population is not necessary to gain control and limit the destruction or transmission of disease. You have several potentially harmful pathogens present in your body. The use of antibiotics, for example, only needs to eliminate the infective pathogen below a certain threshold. The same reasoning holds for the elimination of a mosquito species in the target area to gain the upper hand in the fight against mosquito transmitted diseases. A novel approach to SIT is referred to as "boost SIT". This is where a second control method is used in combination with SIT. It may be the addition of juvenile hormone or the reliance on auto dissemination of which the synergistic result being a reduction in the population of the target species of mosquito.

There are several other mosquito control techniques employed to kill off or at best lower the population density of a mosquito transmitter of disease. The use of sugar baits to attract and kill mosquitoes, spatial repellants, and other small scale niche strategies are employed to control mosquitoes to go along with a new class of insecticides. Traditionally, the use of mosquito fish, copepods, amphibians, and even mosquito larvae of *Toxorhynchitis* that feed upon larvae of other mosquito species have been employed for this non-blood-feeding species to control mosquito populations. Finally, the use of *Wolbachia* and *Bacillus thuringiensis israelensis* bacteria and entomopathogenic fungi that rely upon spore toxicity are being utilized.

The use of mosquito controlling insecticides may not be eliminated in the foreseeable future, therefore they must be economically affordable, environmentally safer, and sustainable in their application. Biological controls are often employed as a synergistic treatment where the sum of the individual controls is greater than the use of a singular application strategy. This is the future of mosquito control strategies.

CHAPTER 25: CLIMATIC EFFECTS

IN the end, the success or failure of an insect family is measured by their diversity and abundance. Albeit the abundance of a given species or family of insect is an elusive number to quantify. The ecology of insects is a fundamental result of the interaction of weather factors and the long-term effect of climate. If you investigate the study of climate through historical events it becomes clear that the present-day rapid change in atmospheric gases, specifically carbon dioxide and methane, have an enormous deleterious effect upon the inner workings of life on earth. Add the anthropomorphic behaviour associated with habitat destruction, leading to biodiversity loss, and the potential increase in novel pathogens are the result. For example, the effects of recent climate interactions and related invasive species have been discovered in Guantanamo Bay, Cuba. A novel mosquito species, *Aedes vittatus*, endemic to the Indian continent and has never been collected anywhere in the Western Hemisphere until recently (18 June 2019). The mosquito is a transmitter of several viruses including chikungunya, Zika, dengue, and yellow fever viruses.

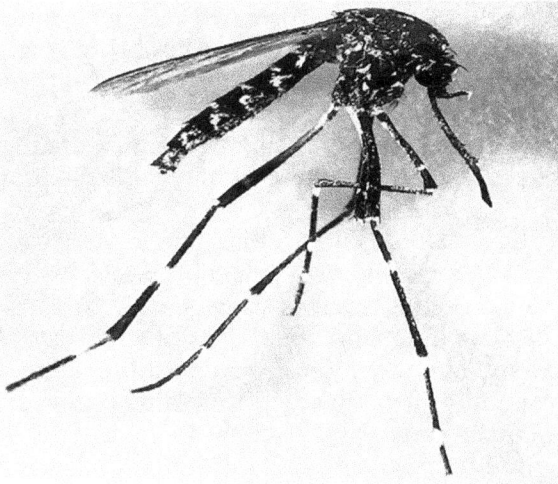

Aedes vittatus. Photo credit: Ben Pagac

Nonetheless, the prevailing attitude of most humans is to view mosquitoes as a pest species. They are well known as vectors of disease to humans and other animals and only serve a minor role as pollinators. It is highly speculative that the mosquitoes serve as an enriched food source for generalists like bats and dragonflies. They are often reported as essential predators, but research suggests this position as an overstated correlation. Certainly, most would agree that mosquitoes do not display the esthetic appeal of the butterfly. Science does not function on the opinions and beliefs but rather develops the methodologies to substantiate empirically the known facts as in this case about the significance or need for mosquitoes. The classic argument debated in classrooms is whether all mosquitoes, or just the disease carriers, or perhaps none of the mosquitoes should be eliminated.

The evolution and existence of any species on earth is a direct result of climatic effects upon the abiotic forces that mold and structure the landscape. I refer you to the geologic timetable that displays the results of abiotic conditions with life's evolutionary path. The resulting effects on life forms can be viewed as having adapted to these forces at work. We all know that the rapid change in our atmosphere, the increase in carbon dioxide and methane gas, has provided the accelerant to rapidly modify and in some cases damage the balance of nature. It is the visual as well as empirical measurements that we can point to. We have all observed the changes displayed in the before and after photos of the melting of glaciers and the shrinking of the polar ice caps. Empirically, we have decades of data showing the reduction of the salinity of our oceans and the projected changes this leads to. There is an effect upon life in the oceans involving plankton, fish, and the whiting of the coral reefs all leading toward a loss of oceanic biodiversity. This is the direct result of rising temperatures. But, you know that. I previously talked about the melting of permafrost and the effect upon the life in the tundra. There are countless of other examples that suggest unequivocally that these abiotic factors work together and ultimately have influence within the mosquito world. I will provide additional examples to this later in the chapter.

There are three main conditions that influence the day to day or even hour to hour weather conditions. Temperature, precipitation, and wind. Weather is a dynamic force that is easily observed, data recorded, and somewhat predictable over short time spans of a few days. Reliance upon satellite pictures and far more sophisticated measurements are used to map weather patterns and provide weather models used in today's forecasting. Climate on the other hand reflects the mean of data usually viewed over a thirty-year period from which some measure of judgement and predictable events can be determined. Climate change is a measure of long-term studies that may involve hundreds, thousands, or even millions of years. Collecting of comparative data provides insight as to what climate in a particular biome was like, what changes occurred, when they occurred, and at what timely rate those changes took place. Climate change today reflects increases in anthropomorphic behavior, primarily the burning of fossil fuels and the destruction of habitat. The increase in carbon dioxide and methane into the atmosphere has accelerated the rate of change to the global climate that requires immediate attention.

It is known that the changes in climate have resulted in changes to global wind patterns, precipitation events, and is increasing temperatures. The overall effect of these factors influences insects including the mosquitoes. In parts of Asia, Japanese encephalitic virus (JEV) infections are an endemic arbovirus where during an outbreak has the potential to kill 35% to 70% of the people. The virus is a member of the *Flavivaridae* family that was once transmitted predominately by *Culex quinquefasciatus*. In 2007, a severe drought eliminated sewage filled watery environments, stagnant and polluted water holes, and dried up natural and artificial containers that held enriched organic solutes. The drought and increase in temperatures changed the ecology of the region to the point that *Cx. quinquefasciatus* was replaced by the latest predominant mosquito, *Cx. triaenierhynchus*, the rice paddy mosquito. This mosquito not only survived the drought but could take advantage from a developmental state to populate and become the dominant mosquito that vectored JEV. The mosquito blood fed upon wild and domesticated pigs that served as reservoirs for the virus to go along with their preference for breeding in cleaner irrigated rice paddies.

Temperature is the most dominant abiotic driver of mosquito life history. Water is also a necessity for mosquito development through all stages except for the adult without which they will not survive. When consideration is given to the metacommunity of drivers that help or harm mosquito existence the need to study all aspects of development with an eye on diseased mosquitoes as transmitters requires science's full attention. All aspects of life driven by weather and climate as the abiotic drivers coupled with the ecological and behavioral aspects of a mosquito's life history reveals areas of study that we often ignore in our quest to control the mosquito yet influence mosquito populations.

Food availability is a bottom-up driver that promotes diversity and abundance for the *Culicidae*. If food is sparse then the consequences become obvious. This can be said about any life form deprived, for whatever reasons, of nutrition. If nutrients are plentiful, then development, increases in abundance, and fecundity are the net result. On the flip side, the top-down driver as a control agent are the mosquito predators. When present in large numbers, the mosquito population of developing larvae decreases substantially. When the predator is absent, mosquito development is often able to prosper.

A study in West Africa provided a model that analyzed the annual mean temperature versus the total annual rainfall. It has been recognized that the increase in temperature accelerates the reproduction of the *Plasmodium* parasite in the mosquito. Furthermore, the increase in temperature increases the longevity of the *Anopheles* mosquito that allows the parasite to complete its developmental cycle within the host mosquito. The greatest incidence of this occurs in the cooler regions where the increase in temperature has the greatest effect upon the host-parasite complex. This correlation follows the known effect of global warming at the higher latitudes in the Northern Hemisphere of the earth. Therefore, higher latitude and higher altitudes play a significant role in the effects of climate change as a key driver of increasing mosquito range, abundance, and vector capacity. As the mean temperatures continue to increase, at what may appear to us as an insignificant 'degree of change', it has been shown repeatedly what a profound effect it has on all forms of life. It is far more influential, these one or two degrees increase in temperature, than we can sense ourselves without the aid of empirical measurement. In its simplest form, look outside at the losses we sense in our forests, glaciers, snow-capped mountains, and the endangerment to all life forms that is accelerating out of control. As far as malaria in Africa goes, the increase in precipitation coupled with increasing temperatures, and the formation of stagnant pools of water has led to the increase in *Anopheles* breeding and the increase in malaria outbreaks.

A final example, reported in the multiple studies of dengue, we know that the incidence of the infection has increased 30-fold in the past 50+ years. Clearly the drivers of climate change and weather patterns has led to an increase in both distribution and incidence of dengue through range expansion on a global level. The *Aedes* mosquito transmitters, *Ae. aegypti* and *Ae. albopictus*, have increased their geographic range expanding into more northerly territories and display a preference for urban and suburban human association increasing the contact ratio of the mosquito and its host. Consequently, several of the poorer income countries carry the burden of social and economic losses.

Throughout Europe, DENV, CHIKV, and ZIKV are of primary importance having been introduced through the increase in travelers to Europe. Invasive mosquitoes capable of transmitting disease have also created concern. *Ae. albopictus*, an invasive mosquito was introduced into Europe in 1979 in Albania. As the temperatures rise, this cold hardy mosquito and its close relative *Ae. aegypti*, is rapidly expanding in a northerly direction. West Nile virus is an expanding mosquito- borne disease in Europe. Climate change has altered the transmission dynamics of the diseased mosquitoes and patterns of pathogenic cycles have occurred in mosquitoes, humans, and other hosts that serve as reservoirs. Reports have suggested an increase in mosquito bite rates from increased rates of mosquito-human contact, an increase in vectoral capacity, shorter diapause times, and an increase in the hatching rates and survivability of the mosquitoes.

Climatic changes caused, in part, from the increase in greenhouse gases have affected the spread of disease by mosquitoes. Variables associated with climate include temperature, precipitation, humidity, and perhaps even wind flow direction and speed. Environmentally, the loss of habitat, land use, and changes in vegetation all lead to an increase in mosquito presence, increases in abundance, vector capacity capabilities, and changes in life history as it relates to seasonality.

CHAPTER 26: FINAL THOUGHTS

I have often thought about and continued to review the process involved in writing my first book. It seems that it had become a lesson in 'frustrated enjoyment'. It was from the beginning an effort to influence young biologists by capturing their imaginative and cognitive interests in the life sciences. The vehicle I employed is of course the role played by the mosquito family. I hope that you have gained insight into the concepts of time and space, the significance of evolution, climate altering manifestations of human behavior, and the importance of pathogens and their ecological, economical, and medical burdens placed upon our lives and the livelihood of other vertebrates. There exists an incredible amount of room for you to enter the realm of science and research. Have I captured your attention? I hope so.

You are very aware of the often-referenced analogy 'the tip of the iceberg'. Well, I think you might want to consider the information in this book as 'the vapors above the iceberg' that surrounds that proverbial tip. Recognize the vastness of learned knowledge and the near infinite knowledge yet to be discovered. After all, you have just begun to educate yourself. The point is best expressed in this statement by Galileo. "You cannot teach a man anything; you can only help him discover it in himself".

I not only wish to have captured your imagination, desire, and thirst for scientific knowledge, but, to add a motivational motif that you may forever remember. I cannot accomplish this task without sharing with you some of the greatest moments of thoughts, messaging, and scholarly advice proffered by so many of the greatest scientists. Remember this advice is meant for you, the emerging scientist. How to think of your role in science is best illustrated with this statement. "It is strange that only extraordinary men make the discoveries, which later appear so easy and simple."- George C. Lichtenberg, 18th century German physicist. You are one of those extraordinary people. Imagination is one of the hallmarks of discovery that lies within all of us. It just must be released. To paraphrase Hubble, you are equipped with five senses, you can explore the universe around you and call the adventure science. Einstein once remarked that imagination (paraphrased) is not limited and is more important than knowledge. In a way only Einstein could convey, he said, "Two things are infinite, the universe and human stupidity: and I am not sure about the universe". A lesson from the past has always stayed with me. Sir Isaac Newton shared his thoughts about discovery, his visionary thoughts, and ultimately his remarkable contributions when he was quoted as saying "If I have seen further it is by standing on the shoulders of giants". I am aware that many of you will experience the feeling of discouragement. It happens to all of us at one time or another with the feeling of disrespect for your work and the incredible ignorance displayed by so many dampens your spirit and drive. Keep these two thoughts in mind when you feel yourself caving to the negativism of society. Isaac Asimov reminds us that "The saddest aspect of life right now is that science gathers knowledge faster than society gathers wisdom". Finally, the words of Neil deGrasse Tyson who is credited with this quote. "The good thing about science is that it's true whether or not you believe in it".

Science is built upon previous knowledge, through an unmistakable sequence of methods, that becomes a fact that was once never given a single thought. My final comment to the reader is simply this. Personal goals set by you should not be pushed too far into your future. I recall someone stating that if it gets too late, those goals, sadly, become never.

BIBLIOGRAPHY

Books:

Burkett-Cadena. 2013. Mosquitoes of the Southeastern United States. The University of Alabama Press, Tuscaloosa, Alabama

Chandler, A. C., C. P. Read. 1961. Introduction to Parasitology. John Wiley & Sons, Inc., 822 pp

Clements, A. N. 2000. The Biology of Mosquitoes, Volume 1,Development, Nutrition, and Reproduction. CABI Publishing, Cambridge, MA

Clements, A. N. 1999. The Biology of Mosquitoes, Volume 2, Sensory Reception and Behavior. CABI Publishing, Cambridge, MA

Clements, A. N. 2012. The Biology of Mosquitoes, Volume 3, Transmission of Viruses and Interactions with Bacteria. CABI Publishing, Cambridge, MA

Darsie, R. F. & R. A. Ward. 2005. Identification and Distribution of the Mosquitoes of North America, North of Mexico. University Press of Florida, Gainsville, FL

Darwin, C. 1859. On the Origin of Species by Natural Selection. J. Murry, London, England

Goddard, J. Infections, Diseases, and Arthropods. 2018. Humans Press 3rd edition, 223 pp.

Grimaldi, D. & M. S. Engel. 2006. Evolution of the Insects. Cambridge University Press, New York, New York

Klowden, M. J. 2007. Physiological Systems in Insects. Elsevier Inc, San Diego, CA

Mayr, E. 2001. What Evolution Is. Basic Books, New York, New York

MacArthur, R. H. & E. O. Wilson. 1967. The Theory of Island Biogeography. Princeton University press, Princeton, NJ

Means, R. G. Mosquitoes of New York. Part I. The Genus Aedes Meigen with Identification Keys to Genera of Culicidae. Bulletin No. 430a. The University of the State of New York, State Science Service, New York State Museum , Albany, New York

Means, R. G. 1987. Mosquitoes of New York. Part II. Genera of Culicidae other than Aedes. Bulletin No. 430b. The University of the State of New York, The State Education Department, State Science Service, New York State Museum, Albany, New York

Montgomery, B. L. 2020. Zika Mozzie Seeker: Citizen Scientists Create Expansive Surveillance Networks for Invasive Aedes Mosquitoes in South East Queensland, Australia. Wingbeats, 31, 3, 5-19

Mullen, G. R., and L. A. Durden. 2009. Medical and Veterinary Entomology, Academic press

Pitkin, R. M. 2008. Whom the Gods Love Die Young. A Modern Medical Perspective on Illnesses that Caused the Early Death of Famous People. Rose Dog Books, Pittsburgh, PA

Service, M. 2012. Medical Entomology for Students. Cambridge University press, New York, New York

Speight, M. R., M. D. Hunter, &A. D. Allen. 2008. Ecology of Insects Concepts and Applications. John Wiley & Sons Ltd, Chichester, West Suffex, UK

Spielman, A. & D'Antonio, M. 2001. Mosquito: The Story of Man's Deadliest Foe. Hyperion, New York, New York

Triplehorn, C. A. & N. F. Johnson. 2005. Borror and DeLong's Introduction to the Study of Insects. Thomas Brooks/Cole, Belmont, MA

Thomas, & J. J. Shepard. 2005. Identification Guide to the Mosquitoes of Connecticut. The Connecticut Agricultural Experiment Station, New Haven, Connecticut. Accessed on July 15, 2020

Zimmer, C. 2015. A Planet of Viruses. The University of Chicago press, Chicago, Illinois, 122 p.

Publications:

Achee, N. L., J. P. Grieco, H. Vatandoust et al. 2019. Alternative strategies for mosquito-borne arbovirus control. PLOS Neglected Tropical Disease, 13 (3) : e0007275. https://doi.org/10.1371/journal.pntd.0007275

Ali, A. & G. Galizic. 2015. Chemosensory Cues for Mosquito Oviposition Site Selection. Journal of Medical Entomology. 52 (2), pp 120-130.

Arab, A., M. C. Jackson, and C. Kongoli. 2014. Modeling the effects of weather and climate on malaria distributions in West Africa. Malaria Journal, 13, 126. https://doi.org/10.1186/1475-2875-13-126

Brakes, M. A. H.,N. A. Honorio, L. P. Lounibos, R. Louenco-de-Oliviera, & S. A. Juliano. 2004. Interspecific competition between two invasive species of container mosquitoes, Aedes aegypti and Aedes albopictus (Dipter: Culicidae), in Brazil. Annals of the Entomological Society of America. 97, pp 130-139.

Baugueras, S., B. Fernandez-Martinez, J. Martinez-dela Puerte, et al. 2020. Environmental diseases, climate change and emerging diseases transmitted by mosquitoes and their vectors in southern Europe: A systematic review. Environmental Research, 191. https://doi.org/10.1016/jenvres.2020.110038

Benelli, G., C. L. Jeffries, and T. Walker. 2016. Biological Control of Mosquito Vectors: Past, Present and Future. Insects, 7 (4), 52

Borkent, A. & D. A. Grimaldi. 2004. The Earliest Fossil Mosquito (Diptera: Culicidae), in Mid-Cretaceous Burmese Amber. Annuls of the Entomology Society of America. 97 (5), pp 882-888.

Briggs, D. E. 2013. A mosquitoes last supper reminds us not to underestimate the fossil record. Proceedings of the National Academy of Science U.S.A. 110 (46), pp 18353-18354.

Ciccozzi, M., S. Pelette, E. Cella et al. 2013. Epidemiological history and phylogeography of West Nile virus lineage 2. Infections, Genetics, and Evolution, 17, 46-50

Courtier-Orgogozo, V., Morizot, B., and Boete, C. 2017. Agricultural pest control with Crispr-based gene drive: time for public debate. Embro reports, 18 (6), 878-880

Dotseth, E. J. & B. A. Harrison. 2016. West Virginia mosquitoes: sequential list by publication, newly found species, corrections, and notes from earlier reads. Journal of Medical Entomology. 32 (3), pp 240-243.

Ebi, K. and J. Neolon. 2016. Dengue in a changing climate. Environmental Research, 151, 115-123 https://doi.org/10.1016/jenvres.2016.07.026

Frank, J. H. 1986. Bromeliads as ovipositional sites for *Wyeomyia* mosquitoe: Form and color influence behaviour. Florida Entomologist. 69, pp 728-742.

Gaston, K. J. & E. Hudson. 1994. Regional Patterns of Diversity and Estimates of Global Insect Species Richness. Biodiversity and Conservation. 3 (6), pp 493-500.

Hansky, I. & M. Gyllenberg.1997. Uniting two general patterns in the distribution of species. Science. 275 (5298), pp397-400.

Harbach, R. E. & D. Greenwalt. 2012. Two Eocene species from of *Culiseta* (Diptera: Culicidae) from the Kishenehn Formation in Montana. Zootaxa. 3530, pp 25-34.

Holmes, E. 2010. The gorilla connection. Nature, 467, 404-405

Huai-Yu, T., P. Bi, B. Cageilles et al. 2015. How environmental conditions impact mosquito ecology and Japanese encephalitis: An eco-epidemiological approach. Environmental International, 79, 7-24.

Jones, C. J. & E. T. Schrieber. 1994. Color and height affects oviposition site preferences of *Toxorhynchitis splendens* and *Toxorhynchitis rutilua rutilus* (Diptera: Culicidae) in the laboratory. Environmental Entomology. 28, pp 130-135.

Ledesma, N. & L. Harrington. 2011. Mosquito Vectors of Dog Heartworm in the United States: Vector Status and Factors Influencing Transmission and Efficiency. Topics in Animal Companion Medicine, 26 (4), 178-185.

Lester, P. J. & A. J. Pike. 2003. Container surface area and water depth influences the population dynamics of the mosquito *Culex pervigilians* (Diptera: Culicidae) and its associated predators in New Zealand. Journal of Vector Ecology. 28, pp 267-274.

Minakawa, N., G. Sonye, & G. Yan. 2005. Relationships between the occurrences of *Anopheles gambia s. l.* (Diptera: Culicidae) and size and stability of larval habitats. Journal of Medical Entomology. 42, pp 295-300.

Mustafa, M. S., V. Rasolgi, S. Jain, & V. Gupta. 2015. Discovery of fifth stereotype of dengue virus (DENV-5): A new public health dilemma in dengue control. Medical Journal Armed Forces India, 71 (1), 67-70.

Nash, D., F. Mostashari, A. Fine et al. 2001. The Outbreak of West Nile Virus Infection in the New York City Area in 1999. New England Journal of Medicine, 44, 1807-1814

Normille, D. 2013. Surprising New Dengue Virus thru a Spanner in Disease Control Effects. Science, 342, 6157, p 415. DOI:10.1126/science.342.6157.415.

Nyasembe, V. O. & B. Torto. 2014. Volatile phytochemicals as mosquitoe semiochemicals. Phytochemistry Letters. 8, 196-201

Onyabe, D. Y. & B. D. Rottberg. 1997. The effect of conspecifics on oviposition site selection and oviposition behavior in *Aedes togoi* (Theobold) (Diptera: Culicidae). The Canadian Entomologist. 129, pp 1173-1176.

Otto, T. D., A. Gilabert, & T. Crellan etal. 2018. Genomes of all known members of a *Plasmodium* subgenus reveal paths to virulent human malaria. Nature Microbiology, 3, 687-697.

Polnar, G. O., T. J. Zavortnik, T. Pike, & P. H. Johnson. 2000. *Palioculicis minutus* (Diptera: Culicidae) n.gen., n. sp., Cretaceous Amber, with a summary of described fossil mosquitoes. ACTH Geologica Hispanica, 35, pp 119-128.

Reisen, W. K., R. P. Meyer, & M. M. Milbey. 1986. Patterns of fructose feeding by *Culex tarsalis* (Diptera: Culicidae). Journal of Medical Entomology. 23, 366-373.

Ruckert, C. & D. E. Gregory. 2018. How do virus-mosquito interactions lead to viral emergence? Trends in Parasitology, 34 (4), 310-321. Doi: 10.1016/j.pt.2017.12.004

Schrama, A., E. E. Gorsich, E. R. Hunting et al. 2018. Eutrophication and predator presence overrule the effects of temperature on mosquito survival and development. PLoS Negl Trop Dis, 12 (3) accessed on 2/6/21 at http://journals.plos.org/plosntds/article?rev=2&id=10.1371/jour-nal.pntd.0006354.

Thomas, M. B. 2018. Biological control of human disease vectors: a per-spective on challenges and opportunities. Bio Control 63, 61-69. https://doi.org/10.1007/s10526-017-9815y

Torrisi, G.J. 2013. Color and Container Size Affect Mosquito (*Aedes trise-riatus*) Oviposition. Northeast Naturalist. 20, pp 363-371.

Torrisi, G. J., W. W. Hoback, J. E. Foster, E. A. Heinrichs, & L. G. Higley. 2015. Predation of Introduced Mosquito Larvae by the Midge *Metriocnemus knabi* in the Phytotelma of the Pitcher Plant *Sarrecenia* purpurea and the Coliniza-tion Following Dry Conditions. Northeast Naturalist. 22 (3), pp513-529.

Ulbert, S. 2019. West Nile virus vaccine- current situation and future di-rections. Human Viruses and Immunotherapeutics, 15, 2337-2342, DOI: 10.1080/21643515.2019.1621149

Van den Hoogen, J. S. Geisen, [...] T. W. Crowther. 2019. Soil nematode abundance and functional group composition at a global scale. Nature, 572, 194-198.

Waheed, I. Bajwa. 2018. A Taxonomic Checklist and Relative Abundance of the Mosquitoes of New York City. Journal of the American Mosqui-to Association. 34 (2), pp 138-142.

Webster, D., Diitrova, K., Holloway, K., Makowski, K., Safronetz, D., & Drebot, M. (2017) California Serogroup Virus Infection Associated with Encephalitis and Cognitive Decline. Canada, 2015. Emerging Infectious Diseases, 23 (8), 1423-1424. https://dx.doi.org/10.3201/eid2308.170239

Yahioka, M., J. Couret, F. Kim, J. McMullan, T. R. Burket, E. M. Dot-son, U. Kltron, & G. M. Vazgog-Prokopec. 2012. Diet and Density Dependent Competition affect Larval Performance and Oviposition Site Selection in the Mosquito *Aedes albopictus* (Diptera: Culicidae). Parasitology Vectors. 5, p 225.

Yanoviak, S. P. 1999. Effects of leaf litter species on macroinvertebrate community structure and mosquito yield in neotropical tree-hole mi-crocosms. Oecologia. 120, pp 147-155.

Yanoviak, S. P. 2001. Container color and location affect macroinverte-brate community structure in artificial tree holes in Panama. Florida Entomologist. 84, pp 315-332.

Yee, D. A., B. Kesavaraju, & S. A. Juliano. 2004. Larval feeding behavior of three co-occurring species of container mosquitoes. Journal of Vec-tor Ecology. 84, pp 315-332.

Zahiri, N. & M. E. Rau. 1998. Oviposition attraction and repellency of *Aedes aegypti* (Diptera: Culicidae) to waters from conspecific larvae subjected to crowding, confinement, starvation, or infection. Journal of Medical Entomology. 35, pp 782-787.

Journal Articles Accessed on the WEB

Alvarez-Hernandez, D. A. & Rivera, S. A. 2017. An Introduction to Vector-Borne Diseases accessed on 11/5/2020 at https://www.researchgate.net/publication/322684544_An_Introduction _to Vector-Borne Disease

American Museum of Natural History. (2008, May 18). Geneticists Trace the Evolution of St. Louis Encephalitis. Science Dailey. Retrieved on December 8, 2020 from https://www.sciencedailey.com/releases/2008/05/080515113308.htm

Barredo, E. 7 M. DeGennaro. 2020. Not Just from Blood: Mosquito Nutrient Acquisition Nectar Sources. Trends in Parasitology, 36 (5), 473-484 accessed on 9/21/2020 at https://www.sciencedigest.com/science/article/pii/S1471492220300404.

Bewick, A., Agusto, F., Calabrese, J. M., Muturi, E. J., & Fagan, W. F. (2016). Epidemiology of LaCrosse Virus Emergence, Appalachia Region, United States. Emerging Infectious Diseases. 22, (11), 1921-1929. Accessed on 12/9/2020 at https://doi.org/10.3201/eid2211.160308.

Briegel, H. 2003. Distinguished achievement Award Presentation at the 2002 Society of Vector Ecology Meeting Physiological bases of mosquito ecology. Accessed on 9/28/2020 at https://www.semanticscholar.org/paper/Distinguished-Achievement -Award-Presentation-at-the-Briegel.pdf

Collins/Center for Disease Control and Prevention (CDC); Image number 9534 accessed on 1/12/2021 at https://www.britanica.com/animal/Aedes-aegypti.

Gruch, J. D., D. Kaemer, & P. Kuru. 2019. Neural Cell Responses upon Exposure to Human Endogenous Retroviruses. Frontiers in Genetics accessed on 11/2020 at https://www.frontiersin.org/articles/10.3389/fgene.2019.00655/full.

Hadley, D. 2020. The Types and Stages of Insect Metamorphosis. Accessed on 8/12/2020 at https://www.thoughtco.com/types-of-insects-metamorphosis-1968347.

Harbach, R. E. 2013. Mosquito Taxonomic Inventory. Accessed at http://mosquito-taxonomic-inventory.info/.

Hershberger, Scott. August 4, 2020. How do Scientists Determine the Ages of Human Ancestors, Fossilized Dinosaurs and Other Organisms. Scientific American. Accessed on 8/4/2020 at https://scientificamerican.com.

Hidden Viruses in the Human Genome. 2016. The Genomics Education Program accessed on 11/5/2020 at https://www.genomiceducation.hee.nhs.ulc/blog/hidden-viruses-in-the-human-genome/.

Kong, X. Q. & C. W. Wu.2010. Mosquito proboscis: An elegant bio microelectromechanical system. Physical Review E, 82 (1). Accessed on 9/19/2020 at https://journals.aps.org/pre/abstract/10.1103/Phys-RevE.82.011910.

Loy, D. E., w. Liu, Y. Li et al. 2017. Out of Africa: origins and evolution of the human malaria parasites Plasmodium falciparum and Plasmodium vivax. International Journal of Parasitology, 47 (2-3), 87-97. Accessed on 10/ 29/2020 at https://www.sciencedirect.com/science/article/pii/S0020751916301229.

Moelling, K. & F. Broeker. 2019. Viruses and Evolution-Viruses first? A Personal Perspective. Frontiers in Microbiology, 10, 523 accessed on 11/5/2020 at https://www.frontiersin.org/articles/10.3389/frieb.2019.00523/full.

Takken, W. & N. O. Verhulst. 2013. Host Preference of Blood-Feeding Mosquitoes. Annul Review of Entomology. 58, 433-453 accessed on 9/21/2020 https://pdfs.semanticscholar.org/46B0/aa950734c2964ad-c7dbbebbof89ebaf5eefb.pdf.

Wessner, D.R. 2010. The Origins of Viruses. Nature Education, 3, 9, p. 37 accessed on 11/4/2020 at https://www.Nature.com/scitable/toxic page/ the-origins-of-viruses-14398218//2020 at https://link.springer.com/chapter /10.1007/978-3-030-02318-8_4.

World Wide Web:

https://agrilife.org/aes/mosquitoes-of-texas accessed on 8/5/2020

https://agrilife.org/aes/mosquitoes-of-texas

https://www.atsu.edu/faculty/chamberlain/arthro.htm

https://wwwbbc.com/future/article/202010115-aedes-vittatus-a-mosqui-to-that-carries-zika-and-dengue

https://www.bu.edu/researchsupport/safety/rshp/agent-informa-tion-sheets/snoeshoe-hare-virus-agent-information-sheet/

https://www.cdc.org/chikungunya/index.html

https://www.cdc.gov/dengue/index.html

https://www.cdc.gov/easternequineencephalitis/index.html

https://www.cdc.gov/globalhealth/newsroom/topics/yellowfever/index.html

https://www.cdc.gov/Jamestown-canyon/index.html

https://www.cdc.gov/lac/index.html

https://www.cdc.gov/malaria/about/disease.html

https://www.cdc.gov/malaria/malaria_worldwide/impact.html

https://www.malariavaccine.org/malaria-and-vaccines/rtss
https://www.cdc.gov/mmwr/volumes/67/wr/mm6717e.htm
https://www.cdc.gov/mosquitoes/about/mosquitoes-in-the-us.html
https://www.cdc.gov/ncezid/dvbd/
https://www.cdc.gov/parasites/lymphaticfilariasis/index.html
https://www.cdc.gov/resources/index.html
http://www.cdc.gov/rubella/vaccinations.html
https://www.cdc.gov/sle/general/qa.html
https://www.cdc.gov/sle/index.html
https://www.cdc.gov/website/statesmap/index.html
https://www.cdc.gov/website/westnile/index.html
https://www.cdc.gov/yellowfever/index.html
https://www.cdc.gov/zikz/prevention/transmission-methods.html
https://cgab.yale.edu/*Center for Genetic analysis of Biodiversity – Yale University*
https://chess.com/*Applications and Investigations in Earth Science. 8th edition*
https://www2.cwhc.ca/wildlife_health_topics/arbovirus/arbossh.php
https://www.denguevirusnet.com/dengue-virus.html
https://www.floridahealth.gov/diseases-and-conditions/mosqui-to-borne-disease/fl-resident-guide-to-mosquito-control-ifas.pdf
http://geology.com/time.htm
https://www.historyofvaccines.org/timeline#EVT_100403
www.health.ny.org
https://www.malaria.com/history-parasites/
https://www.malariasite.com/wars-victims
https://www.mayoclinic.org/disease-conditions/sickle-cell- anemia/symp-toms/syn-20355876
https://www.mayoproceedings.org/article/Soo25-6196(19)30483-S/fulltext
https://pubmed.ncbi.nlm.nih.gov/23100918
https://www.ncbi.nlm.nih.gov/books/NBK470228/
https:///www.ncbi.nlm.nih.gov/books/NBK547682
https://www.ncbi.nlm.nih.gov/pmc/articles/PMC5879000/
https://www.ncbi.nlm.nih.gov/pmc/articles/PMC4297835/ Andreas , T. G., M. C.
at www.sunyjcc.edu/sites/default/files/Mosquitoes-of-Connecticut.pdf
https://www.nps.gov/subjects/geology/time-scale.htm
https://www.newyorker.com/magazine/2019/08/05/how-mosqui-toes-changed-everything

https://www.sciencedirect.com/topics/medicine-and-dentrisry/western-equine-encephalitis-virus

https://www.sciencedirect.com/topics/medicine-and-dentristy/wuchereria-bancrofti

https://www.sciencedirect.com/topics/neuroscience/California-encephalitis-virus

https://web.stanford.edu/group/parasites/ParaSites2006/Lymphatic_filariasis/Disc over.htm

https://www.wbur.org/commonhealth/2019/09/25/eastern-equine-encephalitis-vaccine

https://www.who.int/csr/don/en/

https://www.who.int/data/gho/data/themmes/malaria

https://www.who.int/lymphatic_filariasis/epidemiology/en/

https://www.who.int/malaria/data/en/

https://www.who.int/neglecteddiseases/news/tackling-stigmaization-discrimination-mental-health/en/

https://www.who.int/neglecteddisease/news/treating-more-than-one-billion-peopla-fifth-consecutive-year/en/

https://www.who.int/news-room/fact-sheets/detail/chikungunya

https://www.who.int/news-room/fact-sheets/detail/dengue-and-severe-dengue

https://www.who.int/news-room/fact-sheets/detail/lymphatic-filariasis

https://www.who.int/news-room/fact-sheet/detail/malaria

https://www.who.int/news-room/fact-sheets/detail/vector-borne-diseases

https://www.who.int/news-room/fact-sheets/detail/west-nile-virus

https://www.who.int/news-room/fact-sheet/detail/yellow-fever

https://www.who.int/news-room/fact-sheets/detail/zikz-virus

https://www.who.int/teams/global-malaria-programme

https://www.wrbu.si.edu/vectorspecies/mosquito/forcifer

www.ingramcontent.com/pod-product-compliance
Lightning Source LLC
Chambersburg PA
CBHW050539280326
41933CB00011B/1644